Space Physics Missions Handbook

Office of Space Science
and Applications
Space Physics Division

February 1991

National
Aeronautics and
Space
Administration

Prepared with the assistance of
Science Applications International Corporation
under Contract NASW-4394

This document was prepared for the Space Physics Division of NASA Office of Space Science and Applications by Robert A. Cooper, David H. Burks, and the publications group of the Space Physics Support Division, Science Applications International Corporation. The work in the publications group was directed by Francis P. Glosser and copy-edited by Julie A. Hayne.

The mission handbook provides a summary of current and future flight missions, including those approved and under development, and is a guide for NASA Code-SS scientists and the space physics science community at large.

Requests for further information and suggested corrections and additions for future issues of this handbook should be addressed to:

Robert A. Cooper
Chief Engineer, Space Physics Support Division
SAIC
600 Maryland Avenue SW
Suite 307 West
Washington, DC 20024

Tel: (202) 479-0904
Fax: (202) 488-1122

TABLE OF CONTENTS

Page

1. INTRODUCTION
    1.1 Space Physics 1-1
    1.2 Space Physics Goals 1-1
    1.3 NASA Space Physics Division 1-2
    1.4 NASA Code SS Mission Plan 1-3
    1.5 Purpose of this Handbook 1-3

2. MISSION PLAN
    2.1 Mission Plan 2-1
    2.2 Consensus Scenario Without SEI 2-2
    2.3 Strategy-Implementation Study Consensus Scenario 2-3
    2.4 Potential Cooperative Programs 2-4

3. OPERATIONAL SPACECRAFT
    3.1 Pioneer 10 and 11 .... 3.1-1
    3.2 Interplanetary Monitoring Platform-8 (IMP-8) 3.2-1
    3.3 Voyager 1 and 2 .... 3.3-1
    3.4 International Cometary Explorer (ICE) 3.4-1
    3.5 Dynamics Explorer (DE) 3.5-1
    3.6 Combined Release and Radiation Effects Satellite (CRRES) 3.6-1
    3.7 Ulysses 3.7-1
    3.8 Sounding Rocket Program 3.8-1
    3.9 Scientific Balloon Program 3.9-1

4. APPROVED MISSIONS
    4.1 Solar, Anomalous, and Magnetospheric Explorer (SAMPEX) 4.1-1
    4.2 Solar-A 4.2-1
    4.3 Tethered Satellite System (TSS–1) 4.3-1
    4.4 Wind (GGS) 4.4-1
    4.5 Geotail (COSTR) 4.5-1
    4.6 Spartan-201 .... 4.6-1
    4.7 Waves in Space Plasma (WISP) 4.7-1
    4.8 Polar (GGS) 4.8-1
    4.9 Cluster (COSTR) 4.9-1
    4.10 Solar and Heliospheric Observatory (SOHO) 4.10-1
    4.11 Energetic Heavy Ion Composition (EHIC) 4.11-1
    4.12 Atmospheric Laboratory for Applications and Science
         (Atlas-1) 4.12-1

    4.13   Fast Auroral Snapshot Explorer (FAST) 4.13-1
    4.14   Advanced Composition Explorer (ACE) 4.14-1

5. PLANNED MISSIONS

    5.1   Neutral Environment With Plasma Interaction Monitoring System
           (NEWPIMS) 5.1-1
    5.2   Heavy Nuclei Collector (HNC) 5.2-1
    5.3   Orbiting Solar Laboratory (OSL) 5.3-1
    5.4   Ultra High Resolution Extreme Ultra Violet
           Spectroheliograph (UHRXS) 5.4-1
    5.5   Astromag 5.5-1

6. CANDIDATE FUTURE MISSIONS

    6.1   Lunar Calorimeter 6.1-1
    6.2   Neutrino Astrophysics 6.2-1
    6.3   Tethered Multiprobe 6.3-1
    6.4   ENA/EUV Imager 6.4-1
    6.5   Magnetopause Mapper (Ionosonde) 6.5-1
    6.6   Lunar Solar Observatory 6.6-1
    6.7   ITM Coupler 6.7-1
    6.8   Mesosphere Structure, Dynamics, and Chemistry 6.8-1
    6.9   Imaging Super Cluster 6.9-1
    6.10   Grand Tour Cluster 6.10-1
    6.11   High-Energy Solar Physics (HESP) 6.11-1
    6.12   Solar Probe 6.12-1
    6.13   Polar Heliosphere Probe 6.13-1
    6.14   Interstellar Probe 6.14-1
    6.15   Time Dependent Global Electrodynamics 6.15-1
    6.16   Mercury Orbiter 6.16-1
    6.17   Mars Aeronomy Observer 6.17-1
    6.18   Global Solar Mission 6.18-1

APPENDICES:

    A.   Launch Sites A-1
    B.   Launch Vehicle Performance B-1
    C.   Acronyms C-1

LIST OF FULL- AND HALF-PAGE FIGURES

Page

2. MISSION PLAN

Mission plan 2-1

Consensus scenario without SEI 2-2

Strategy-Implementation Study consensus scenario 2-3

Potential cooperative missions 2-4

3. OPERATIONAL SPACECRAFT

Pioneer 10 and 11
- Pioneer orbits 3.1-5
- Internal arrangement of spacecraft equipment compartment 3.1-6
- Internal view of major spacecraft subsystems 3.1-7
- External view of the spacecraft 3.1-8

Interplanetary Monitoring Platform-8 (IMP-8)
- IMP-8 orbit 3.2-4
- IMP appendages 3.2-5
- IMP structural launch configuration 3.2-5
- IMP-8 spacecraft 3.2-6

Voyager 1 and 2.
- Voyager spacecraft 3.3-4
- Expected profiles of RTG power output aboard Voyager spacecraft 3.3-5
- Spiral solar current sheet shown out to 20 AU 3.3-6
- Voyager orbits 3.3-7
- Hypothetical heliosphere model, projected into solar equator plane 3.3-8

International Cometary Explorer (ICE)
- ICE-configuration 3.4-4
- ICE attitude and orbital control system pictorial 3.4-5
- Top view of ICE with attitude sensor data 3.4-6
- The ICE Earth-return trajectory (1983 to 2014) 3.4-7
- ICE heliocentric trajectory 3.4-8

Dynamics Explorer (DE)
- Dynamic Explorer-1 .... 3.5-3
- Dynamics Explorer 3.5-4
- DE-1 orbit: 4.0 to 6.5 years (1.08 x 4.66 R_e fixed orbit) 3.5-5
- First 3.5 years of DE-1 orbit (1.08 x 4.66 R_e fixed orbit) 3.5-6

Combined Release and Radiation Effects Satellite (CRRES)
- CRRES spacecraft arrangement 3.6-5
- Perspective cutaway view of CRRES 3.6-6
- Perspective cutaway view of CRRES 3.6-7
- Perspective cutaway view of CRRES 3.6-8
- CRRES in launch configuration 3.6-9

  CRRES launch profile 3.6-10
  CRRES satellite chemical release experiments 3.6-11

 Ulysses
  Ulysses spacecraft in-flight configuration 3.7-4
  Ulysses 3.7-5
  Ulysses spacecraft configuration 3.7-6
  Typical Ulysses spacecraft trajectory 3.7-7

Sounding Rocket Program
 NASA sounding rockets 3.8-3

Scientific Balloon Program
 NASA balloon size 3.9-3
 View from the South Pole of the trajectory of balloons 3.9-4
 Schematic diagram of long-duration flight systems 3.9-4
 Balloon payloads and volumes 3.9-5

4. APPROVED MISSIONS
 Solar-A
  Solar-A spacecraft 4.2-3

 Wind (GGS)
  Concept of Wind spacecraft configuration 4.4-4
  Wind orbit 4.4-5

 Geotail (COSTR)
  Geotail in-flight configuration 4.5-4
  Geotail distant tail orbit (working model) 4.5-5
  Plasma sheet model and Geotail's near-tail orbit 4.5-6

 Spartan-201
  Spartan-201 .... 4.6-3
  Spartan spacecraft configuration: all up configuration 4.6-4
  Spartan Release Engine Mechanism (REM): REM base and subsystems 4.6-5
  Spartan 2 .... 4.6-6

 Waves in Space Plasma (WISP)
  WISP 4.7-3

 Polar (GGS)
  Concept of Polar spacecraft configuration 4.8-4
  Polar's orbit 4.8-5

 Cluster (COSTR)
  One of the four identical Cluster spacecraft in flight configuration 4.9-3
  Cluster orbit at six month intervals 4.9-4

 Solar and Heliospheric Observatory (SOHO)
  SOHO insertion trajectory and halo orbit 4.10-4
  SOHO semi-exploded view 4.10-5

 Energetic Heavy Ion Composition (EHIC)
  EHIC large telescope 4.11-3

Atmospheric Laboratory for Applications and Science (Atlas-1)
    Atlas-1 .... 4.12-5

Fast Auroral Snapshot Explorer (FAST)
    The FAST spacecraft 4.13-3

Advanced Composition Explorer (ACE)
    ACE 4.14-4

5. PLANNED MISSIONS

Astromag
    Instruments on Astromag: SCIN/MAGIC 5.5-5
    Instruments on Astromag: Wizard and LISA 5.5-6

6. CANDIDATE FUTURE MISSIONS

Lunar Calorimeter
    Lunar Calorimeter 6.1-3

Magnetopause Mapper (Ionosonde)
    Lunar-based Magnetosphere Sounder Array concept 6.5-3

ITM Coupler
    Coupling and dynamics in the cascading of particles, fields, and waves 6.7-9
    ITM Coupler orbit configuration 6.7-10
    Ionosphere, Thermosphere, Mesosphere Coupler 6.7-11

Mesosphere Structure, Dynamics, and Chemistry
    Mesosphere Structure, Dynamics, and Chemistry 6.8-7

Imaging Super Cluster
    Imaging Super Cluster 6.9-11

High-Energy Solar Physics (HESP)
    High-Energy Solar Physics 6.11-8
    High-Energy Solar Physics Mission in orbit 6.11-9
    High-Energy Solar Physics Mission at launch 6.11-10
    HESP Mission 6.11-11
    HESP Delta II configuration 6.11-12
    Schematic illustration of the basic components of HEISPEC 6.11-13
    Spectral resolution of the HEISPEC 2-segment HPGe detectors 6.11-14
    Rotating modulation collimators and Fourier processing 6.11-15

Solar Probe
    Solar Probe orbit 6.12-6
    Solar Probe technical and programmatic review 6.12-7
    Solar Probe near perihelion reference trajectory 6.12-8
    Solar Probe spacecraft 6.12-9
    Baseline 3-axis stabilized spacecraft configuration—top view 6.12-10
    Baseline 3-axis stabilized spacecraft configuration—side view 6.12-11
    ΔV-EJGA spacecraft launch configuration on the Titan IV/Centaur vehicle 6.12-12
    Solar Probe instruments 6.12-13

Polar Heliosphere Probe
    1988 Helio-Polar Probe to 20 AU 6.13-6
    Polar Heliospheric Probe configuration and deep space ΔV 6.13-7

Interstellar Probe
    Interstellar Probe 2011 launch trajectory 6.14-5

Mercury Orbiter
    Mercury Orbiter spacecraft system internal configuration 6.16-4
    2002 Earth to Mercury trajectory (E-VV-MM-M) 6.16-5
    Mercury phase orbital design 6.16-6

Mars Aeronomy Observer
    Mars Orbiter transfer orbit 6.17-5
    Mars Aeronomy Orbiter mission configuration concept 6.17-6

Global Solar Mission
    In-Ecliptic Network—Solar Orbiter pre-circularization configuration 6.18-6
    In-Ecliptic Network trajectory 6.18-7

APPENDICES:
A.   Launch Sites
    Launch sites A-1
    Sounding rocket launch sites A-3

B.   Launch Vehicle Performance
    Launch vehicle performance B-1
    NASA balloon performance load-altitude curves B-4
    NASA sounding rocket performance

Section 1
Introduction

Section 1 Introduction

1.1 Space Physics

Space physics is, to a great extent, the study of naturally occurring plasmas. Approximately 99% of all the matter in the Universe exists in the form of plasmas of many different types; including the partially-ionized, relatively cool plasmas of planetary ionospheres, the million-degree plasmas found in the solar corona, the solar-wind plasmas, the planetary magnetospheres, and the highly relativistic galactic cosmic-ray plasma. Not only do these plasmas have very different physical scales, but each has phenomena occurring on a wide range of scales within it. The challenge for space physics is to arrive at an integrated view that relates large-scale and small-scale plasma phenomena by drawing upon concepts from the frontier of modern scientific research.

In particular the Sun and the heliosphere—the nearest star and its sphere of influence—harbor a large number of fundamental questions that are of consequence not only for the solar system, but also for astrophysics as a whole. Mankind now has the intellectual curiosity and technical capability to investigate the many basic and interconnected questions regarding the internal structure of the Sun, the heating of the corona, and coronal expansion into the fast and slow streams of the solar wind.

Exploration of the Earth's nearby space environment has revealed a dynamic and complex system of plasmas interacting with the magnetic fields and electric currents surrounding our planet. This region, comprising the magnetized solar-wind plasma plus the perturbation in the heliosphere caused by the presence of the magnetic Earth, is the region defined as Geospace. Solar influence shapes and links the three major regions of geospace: the magnetosphere, the ionosphere, and the Earth's neutral atmosphere, where life exists.

Thus space physics can be characterized as the study of the heliosphere *as one system*; that is, of the Sun and solar wind, and their interactions with the upper atmospheres, ionospheres, and magnetospheres of the planets and comets, with energetic particles, and with the interstellar medium.

1.2 Space Physics Goals

Over the next twenty years space physicists will continue to explore the space environment of the Earth and Sun to achieve a new kind of scientific understanding. The new frontiers will be the inner and the outer limits of the solar system; with the Sun as a variable star at its center and the interstellar medium—the frontier of the galaxy—on the outside.

Once the Sun and interstellar medium have been studied *in situ*, it will be possible to start the integrated study of the entire heliosphere as one interacting system. It is a subject that is breathtaking both in the scope of the physics involved, and in its potential application to routine human activities on Earth, including the operation of near-Earth spacecraft.

The knowledge gained by this research program will be critically important to understanding the effects of energetic particles and solar variability upon the Earth's environment and the human exploration of space.

1.3 NASA Space Physics Division

The Space Physics Division (Code SS) is part of the NASA Office of Space Science and Applications (OSSA), and has discipline branches in the following scientific areas:

- **Cosmic and Heliospheric Physics**

    This branch studies the origin and evolution of galactic cosmic rays and solar-system material, acceleration processes, galactic confinement processes, and the transport of energy, plasmas, and magnetic fields in the heliosphere and beyond. Also studied are the wave-particle and plasma-field interactions of the solar wind, including interplanetary shocks.

- **Solar Physics**

    This branch studies the interior, photosphere, chromosphere, transition region, and corona of the Sun, including the generation, storage, and release of solar flare energy. The Solar Physics branch pursues research in nuclear processes, atomic and molecular collisions, magnetohydrodynamics, magnetically confined plasmas, and comparative stellar studies. Helioseismology and studies of solar activity constitute major components of the program.

- **Magnetospheric Physics**

    This branch studies the global structure and microphysical dynamics of magnetospheres, and the interactions of magnetospheres and other obstacles with space plasmas. Research emphases are on planetary magnetospheres, satellite-plasma interactions, and cometary environments.

- **Ionospheric, Thermospheric, and Mesospheric Physics**

    This branch studies the upper atmospheres, ionospheres, and auroral processes of the Earth and other planets, including current-generation and critical-velocity phenomena. The Ionospheric, Thermospheric, and Mesospheric Physics branch aims to understand the formation, structure, coupling, and dynamics of these systems.

The Space Physics Division supports investigations of the origin, evolution, and interaction of particulate matter and electromagnetic fields in a wide variety of space plasmas, including the energy flow and particle transport from the solar surface through the geospace environment to the Earth's upper atmosphere.

Observations, theory, modeling, simulations, laboratory studies, interactive data analysis, instrument development, and active experiments are all important aspects of the space physics research program. Observations are made from a variety of platforms including the Earth itself, high-altitude balloons, sounding rockets, satellites, and interplanetary spacecraft.

Program management for NASA's suborbital programs, both sounding rockets and balloons, resides within the Space Physics Division. The Sounding Rocket and Large Balloon Programs are tracked as two individual programs, each with multiple associated launches. These programs are low-cost, quick-response efforts that at present provide approximately 40–50 flight opportunities per year to space scientists involved in the disciplines of upper atmosphere, plasma physics, solar physics, planetary atmosphere, galactic astronomy, high-energy astrophysics, and micro-gravity research. This handbook has not addressed the individual suborbital missions within these programs, but rather provides an overview of capabilities in Appendix B.

1.4 NASA Code SS Mission Plan

The diversity of science objectives within the disciplines of space physics demands a broad mix of missions. This is approached ideally through a coordinated set of major missions, which form the backbone of the science, combined with Explorer-class spacecraft aimed at specific, focused scientific problems. Thus there is a need for major missions to perform detailed studies on a global scale, and explore previously unexamined regions (*e.g.*, the environment near the Sun); for moderate missions (Explorer-class) to attack specific, detailed problems; for quick response techniques such as balloons, sounding rockets, and experiments of opportunity that are best accommodated on the Shuttle; and for facility-class instruments that are developed for shuttle and space station, but may evolve towards various space platforms.

A significant thrust in each of the space physics discipline areas is essential to maintain progress in the field, and an adequate level of flight activity must be maintained throughout the program to assure continuity and to involve the science community. In order to achieve this goal, and to identify candidate future missions that will accomplish forefront research, NASA Code SS (Space Physics Division) has undertaken a Strategy-Implementation Study including two Workshops. The first Workshop was held in Baltimore, MD during January 1990, followed by a second held in Bethesda, MD during June 1990. These Workshops also updated the strategy and design of a prioritized implementation plan for inclusion in the OSSA Strategic Plan.

Candidate missions were selected at the first workshop on the basis of the priority of the science questions which each can address, technological readiness, programmatic feasibility, and the necessity to fit within a limited budget.

1.5 Purpose of this Handbook

The purpose of this handbook is to provide background data on current, approved, and planned missions, including a summary of the recommended candidate future missions that emerged from the two workshops. As the priorities of science questions to be addressed are agreed upon, and technical assessments are developed, this report will be updated to reflect the developing implementation plan.

Section 2
Mission Plan

Section 2 Mission Plan

2.1 Mission Plan

The following pages reflect the current Space Physics Division (SPD) mission plan. Data concerning the future missions in the consensus scenario emerged from the recent Strategy Implementation Workshops, and continue to change as more detailed studies are performed.

The SPD mission plan covers the period 1995–2010 and was constructed through the grass-roots involvement of the science community via the conduct of open workshops in the four broad discipline areas of Solar, Cosmic and Heliospheric, Magnetospheric, and Ionospheric-Thermosphere-Mesospheric Physics. The products resulting from the workshops, together with theory, were integrated into an overall SPD program plan that has combined the highest-priority elements from each area into a timeline that achieves discipline balance, a mix of major, moderate, and intermediate missions, and a roughly constant funding profile of about $600M per year. Major missions (> $1B) include the Solar Probe (currently in the OSSA Strategic Plan) and the Ionosphere-Thermosphere-Mesosphere Coupler; moderate missions (~ $500M) are the Mercury Orbiter, High Energy Solar Physics, and Grand Tour Cluster; and intermediate missions (≤ $200M) include Auroral Cluster, Ultra Heavy Cosmic Ray (UHCR), Inner Magnetosphere Imager, Solar Probe Coronal Companion, and Thermosphere, Ionosphere, Mesosphere Energetics and Dynamics (TIMED). The decision rules and boundary conditions adopted for the construction of the plan will be described and possible options and fall-back positions will be discussed. The latter are particularly pertinent in view of the current budgeting uncertainty surrounding the overall NASA budget.

In the process of designing this program the study group, in addition to assuring that the program's three themes for Space Physics were satisfied (*i.e.*, exploring the solar atmosphere and the inner heliosphere, understanding the Sun and its effects upon the Earth's magnetosphere and upper atmosphere; exploring to the frontiers of the heliosphere and interstellar space), evolved the following guidelines: (1) to advance understanding in each of the domains of space physics (*i.e.*, solar physics, magnetospheric physics, ionospheric-thermosphere-mesospheric physics, and cosmic and heliospheric physics) and the chain of interaction among them, a broad scientific attack is necessary. This requires a major thrust in each of the separate disciplines identified above. (2) There is a need for major missions for studies on a global scale, moderate missions for more specific problems, and intermediate missions for sharply focused, constrained studies that attack specific scientific problems. (3) It is essential to establish the level of flight activity necessary to address significant scientific issues. (4) The plan must be accommodated within a realistic budget that can remain stable over a period of several years and allow the planning and analysis essential to a successful research program.

The diagrams on the following pages illustrate the relationship and phasing of the future missions to other major NASA programs, and the current approved and candidate future missions flight schedule.

Section 2 Mission Plan

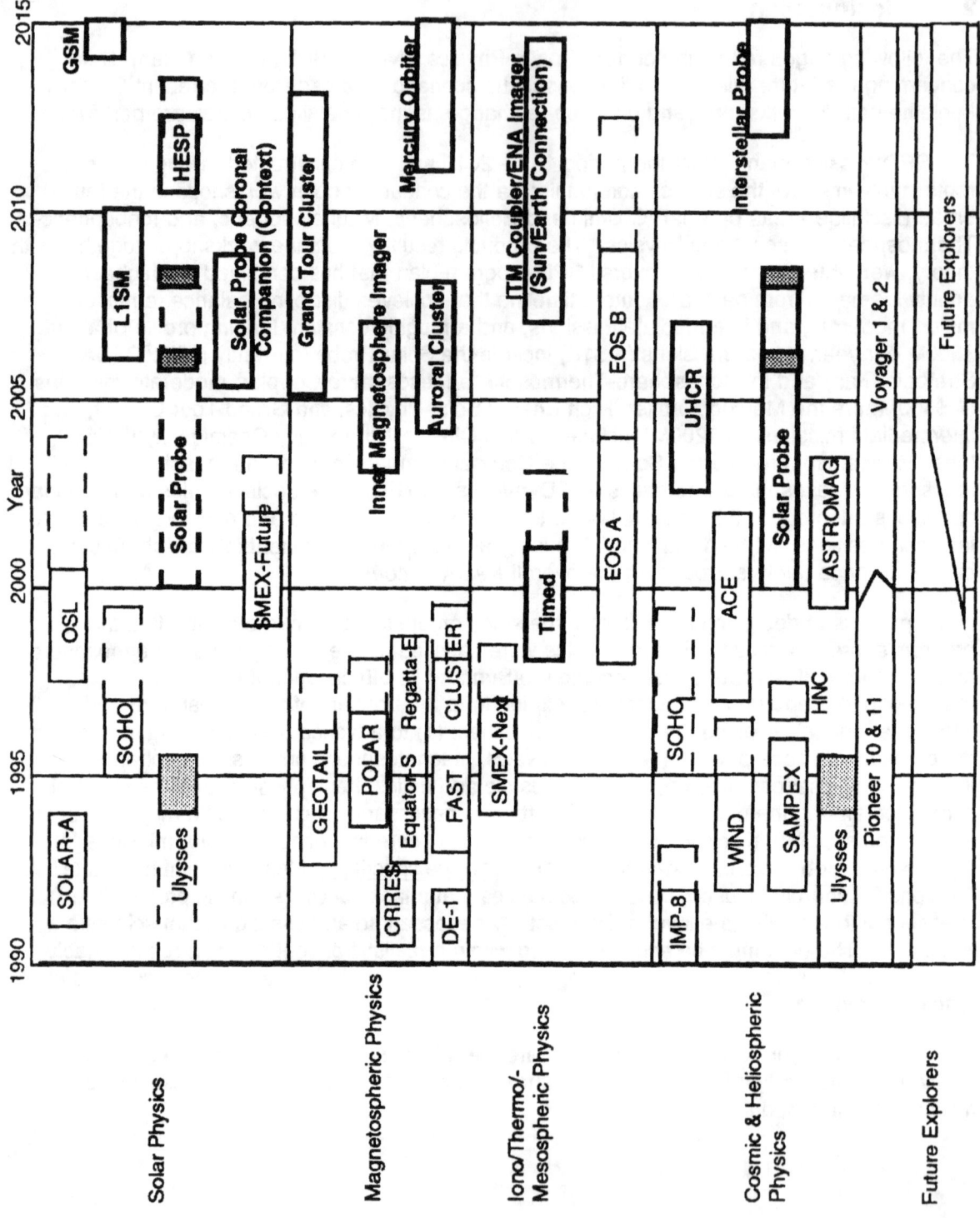

2-2

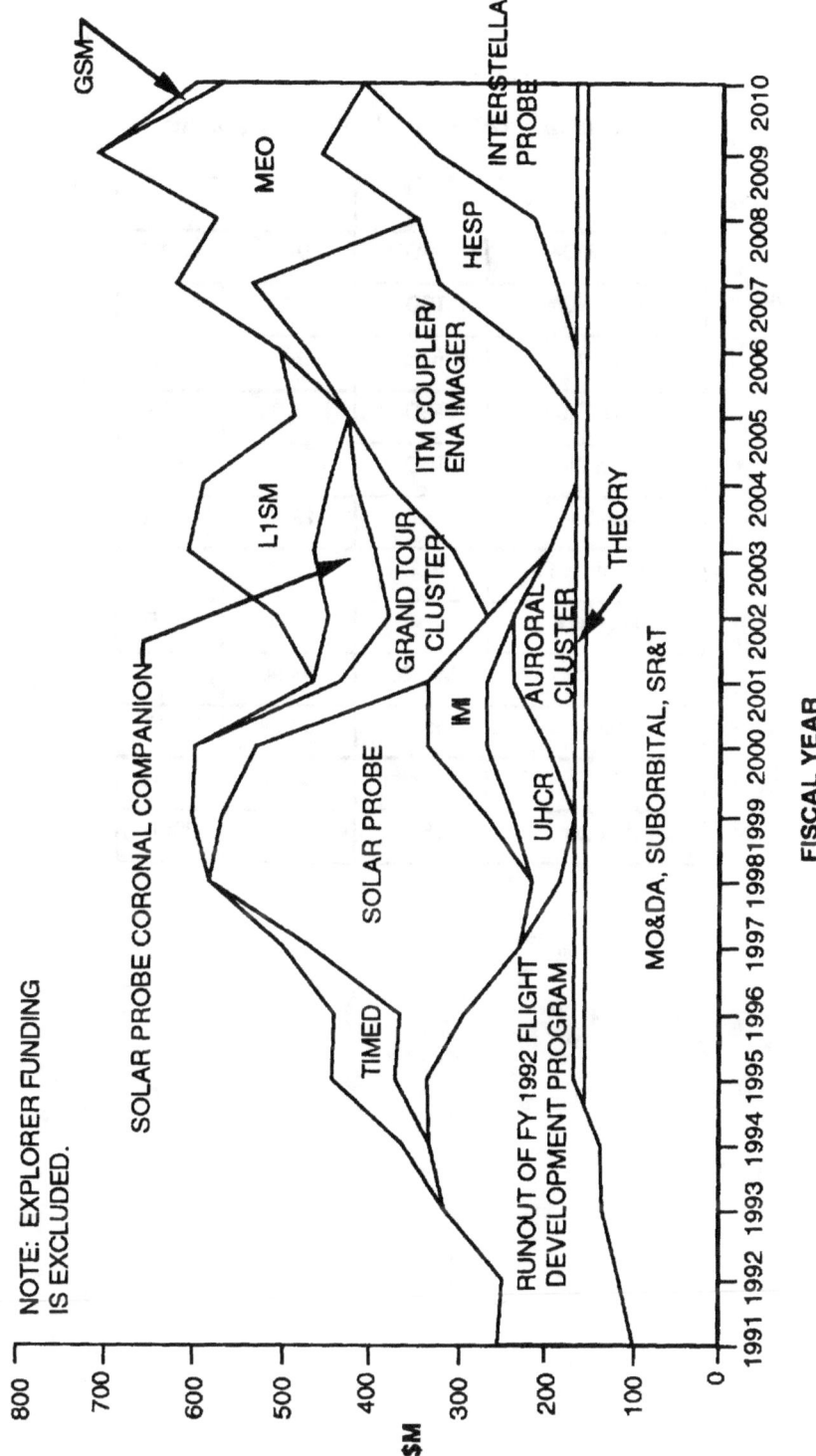

Space Physics Strategy-Implementation Study consensus scenario

Section 2 Mission Plan

Potential Cooperative Programs

MISSION	DURATION	LAUNCH DATA DATE	SITE
CERN/HNC/Mir	TBD	1991	USSR
APEX	6 months	Jun-90	USSR
Regatta Equator	TBD	Dec-93	USSR
Regatta Cluster	TBD	Dec-95	USSR
Interball	TBD	1991	USSR
Coronas I	TBD	1992	USSR
Coronas F	TBD	1994	USSR
PHOBOS	Mar-89	Jul-88	USSR
Active	6 months	Aug-89	USSR
AWE	6 months	Aug-90	USSR
Photon	TBD	TBD	USSR

Section 3
Operational Spacecraft

3.1 Pioneer 10 and 11

Target: Interstellar space

Orbit: Solar system escape trajectory

	Pioneer 10	Pioneer 11
V_α	2.384 AU/yr.	2.214 AU/yr
Ecliptic Lat.	2.909°	12.596°
Ecliptic Long.	83.368°	291.268°

Mission Duration: To RTG expiration

Mission Class: Moderate

Mass: 259 kg

Launch Vehicle: Atlas/Centaur
2 March 1972 (Pioneer 10)
5 April 1973 (Pioneer 11)

Theme: Cosmic and heliospheric physics

Science Objectives:

Space Physics Division will assume management responsibilities for the Pioneer 10 and 11 spacecraft during fiscal 1992. The goals of these two spacecraft over the following decade will be as follows:

- To search for the heliospheric boundary with interstellar space, and study the large-scale electrodynamic structure of the solar plasma and magnetic field.

- To measure the intensity and composition of the galactic cosmic radiation, and study the radial gradient of cosmic ray intensity and its dependence on solar activity.

- To search for gravitational radiation (gravity waves).

Spacecraft:

    Type: Pioneer class, spin stabilized
    Special Features: Four RTGs for primary power
    Special Requirements: TBD

Section 3 Operational Spacecraft

Pioneer is a highly reliable spacecraft of relatively simple design in which many of the components and subsystems have already demonstrated successful performance on earlier missions. It has a thermally-controlled equipment compartment with two sections, one hexagonally shaped, and containing electronic units and the propellant tank, and the other a bay containing most of the scientific sensors and their associated electronics (the magnetometer sensor and two meteoroid detectors are external). Forward of the equipment compartment is a 2.7 m diameter parabolic reflector for the high-gain antenna. Mounted on a tripod structure forward of the reflector are the medium-gain antenna and the feed for the high-gain antenna. Three appendages are stowed within a 2.7 m cylindrical envelope at launch; they are shown in their deployed positions attained within an hour after launch. Two pairs of Radioisotope Thermoelectric Generators (RTGs) are extended approximately 1.8 m at 120 degrees spacing. The RTGs were retained in a stowed position for launch net to the equipment compartment and under the antenna reflector. The magnetometer sensor is located on the end of a long folding boom which, in the deployed condition, extends 5.2 m radially from the instrument side of the equipment compartment.

Six 4 N hydrazine thrusters are located in three clusters near the perimeter of the 2.7 m reflector. Two pairs of thrusters are aligned parallel to the spin axis for precession and velocity correction maneuvers; two thrust tangentially for spin control. Other external features include a mast-mounted omnidirectional antenna directed aft, and a Sun sensor mounted near one of the thruster assemblies which determines the spacecraft's position in the spin cycle. Two large light shields are associated with the stellar-reference assembly, and with an optical asteroid/meteoroid detector.

Instruments:

There are ten scientific instruments on board the two nearly identical spacecraft:

- *Helium Vector Magnetometer (JPL/HVM).* The JPL/HVM measures the vector magnetic field in the heliosphere, with selectable operating ranges from ± 4.0 gamma to 1.41 gauss.

- *Plasma Analyzer (ARC/PA).* The ARC/PA is an electrostatic energy per unit charge (e/q) spectrometer capable of measuring the flux as a function of E/q and incident direction of positive ions and electrons. This instrument is capable of determining incident plasma distribution parameters over the energy range of 100–18,000 eV for protons and approximately 1–500 eV for electrons. It covers the dynamic range for charged particle fluxes from 1×10^2 to 3×10^8 cm^{-1} s^1 and is capable of resolving proton temperature down to at least the 2×10^3K level.

- *Charged Particle Instrument (UC/CPI).* The UC/CPI separately identifies individual nuclei, including protons and helium nuclei, through to the higher mass nuclei up to oxygen and measures the energy and differential flux of these particles over the range from 0.5 to 500 MeV/nucleon. Integral fluxes of nuclei with energies greater than 500 MeV/nucleon from protons through iron are also measured. Electron spectra are measured from 3 to 30 MeV.

- *Geiger Tube Telescope (UI/GTT)*. This instrument utilizes seven GM tubes as elementary detectors. Three tubes are arranged in an array parallel to the X–Y plane of the spacecraft to form a telescope for penetrating particles (Ep>70 MeV) moving in the +Z or -Z (spacecraft rotational) axis direction. The useful dynamic range extends from 0.2 to 1×10^6 counts per second for individual tubes. Three other detectors are arranged in a triangular array and fully enclosed in a 7.2 g/cm^{-2} shield of lead to form a shower detector. A final detector is configured as a scatter detector, admitting large-energy electrons Ee>0.06 MeV) but discriminating against protons (Ep>20 MeV).

- *Trapped Radiation Detector (UCS/TRD)*. This is a Cherenkov detector which measures the gradient and transport properties of very high-energy cosmic rays. It counts galactic cosmic rays in the range above 500 MeV/nucleon, and measures anisotropies perpendicular to the spacecraft spin axis. Thus the radial gradient of the galactic cosmic rays and their modulations caused by the solar radiation are being observed.

- *Cosmic Ray Telescope (GSFC/CRT)*. The instrument comprises three solid state telescopes. The high-energy telescope is a three-element linear array operating in two modes: penetrating and stopping. For penetrating particles, differential energy spectra are obtained for He and H_2 from 50–800 MeV/nucleon. The stopping particle mode covers the range from 22–50 MeV. The low-energy telescope I responds to protons and heavier nuclei from 3 to 22 MeV/nucleon, providing both energy spectra and angular distribution over this range. Low-energy telescope II is designed primarily to study solar radiation. It will stop electrons in the 50–150 keV range and protons in the 50 keV^{-3} MeV range, and will respond to electrons in the interval 150 keV^{-1} MeV and protons from 3 to 20 MeV.

- *Ultraviolet Photometer (USC/UV)*. The USC/UV is a two channel UV photometer operating in the 200–1400 Angstrom range, with a field of views (FWHM) of 1.15° x 9.3°. An aluminum filter in conjunction with a channeltron sensor provides hydrogen lyman-alpha data at 1216 Angstrom, and a lithium fluoride target cathode with a second channeltron sensor provides helium data at 584 Angstrom.

- *Imaging Photopolarimetry (UA/IPP)*. This experiment consists of an optical telescope with a 2.5 cm aperture and 8.6 cm focal length, providing an instantaneous field of view of 40x40 mrad. A Wollaston prism splits the image into two orthogonal polarized beams which are filtered to two color channels: 3900–4900 Angstrom (blue) and 5900–7000 Angstrom (red).

- *Meteoroid Detector (LaRC/MD)*. This micro-meteoroid detector makes *in situ* measurements of solid particle population in the 10^{-8}g mass range and larger using penetration cells attached to the exterior of the spacecraft.

- *Radiometric Science*. Doppler tracking of the spin-stabilized spacecraft is used to search for gravitational radiation.

Mission Strategy:

Pioneer 10 is at a heliocentric radius of 50.4 AU (Jan 91), and is traveling towards the tail region of the heliosphere in a direction opposite to that of the Sun's motion through the galaxy at about 2.7 AU per year. Pioneer 11 is 32 AU from the Sun (Jan 91), and travelling in the bow-shock direction of the heliosphere at about 2.5 AU per year.

The Pioneer spacecraft are now probing unexplored regions of deep space, and will provide the first *in situ* measurements of the large-scale plasma and magnetic-field structure of a type G star (the Sun). There is a reasonable expectation that the Pioneer spacecraft will penetrate the heliospheric boundary before their RTG power output drops below the level of spacecraft requirements, and one of the first indications of this penetration may be the cessation of the 26-day rotation-induced modulation of cosmic rays <100 MeV/nucleon.

The projected amount of electrical power from the RTGs—supplemented by radioisotope heating units—will be adequate to maintain the temperatures of spacecraft subsystems and the instrument bay at a comfortable level and to supply the necessary electronic power for the spacecraft and instruments until the mid-1990s. The supply of fuel for the attitude control system is more than adequate for this time period.

Enabling Technology Development: None

Points of Contact:

Program Manager:	Jim Willett (202) 453-1514
Program Scientist:	Vernon Jones (202) 453-1514
NASA Center:	ARC
Project Manager:	R. O. Fimmel
Project Scientist:	Palmer Dyal

Section 3 Operational Spacecraft

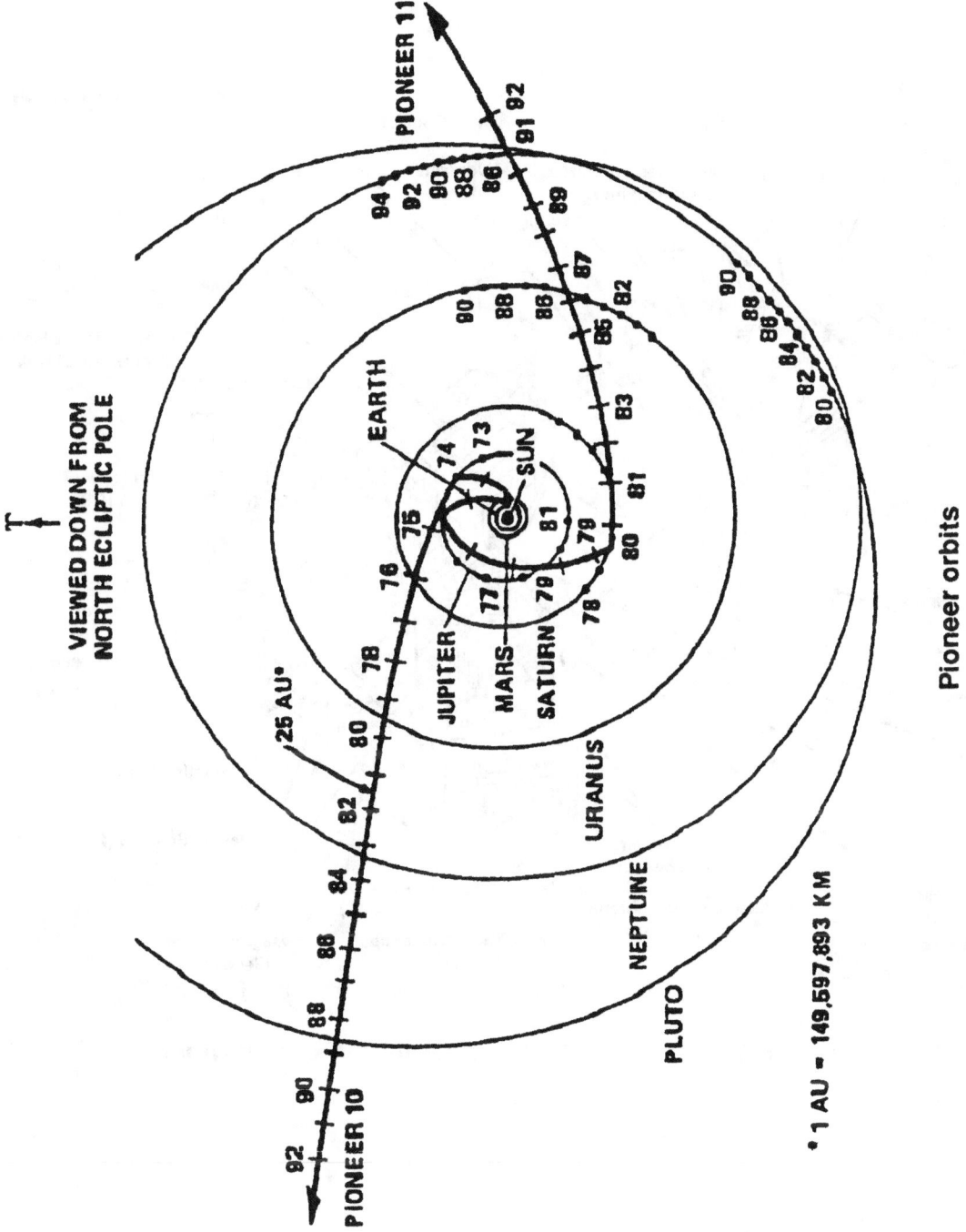

Pioneer orbits

Section 3 Operational Spacecraft

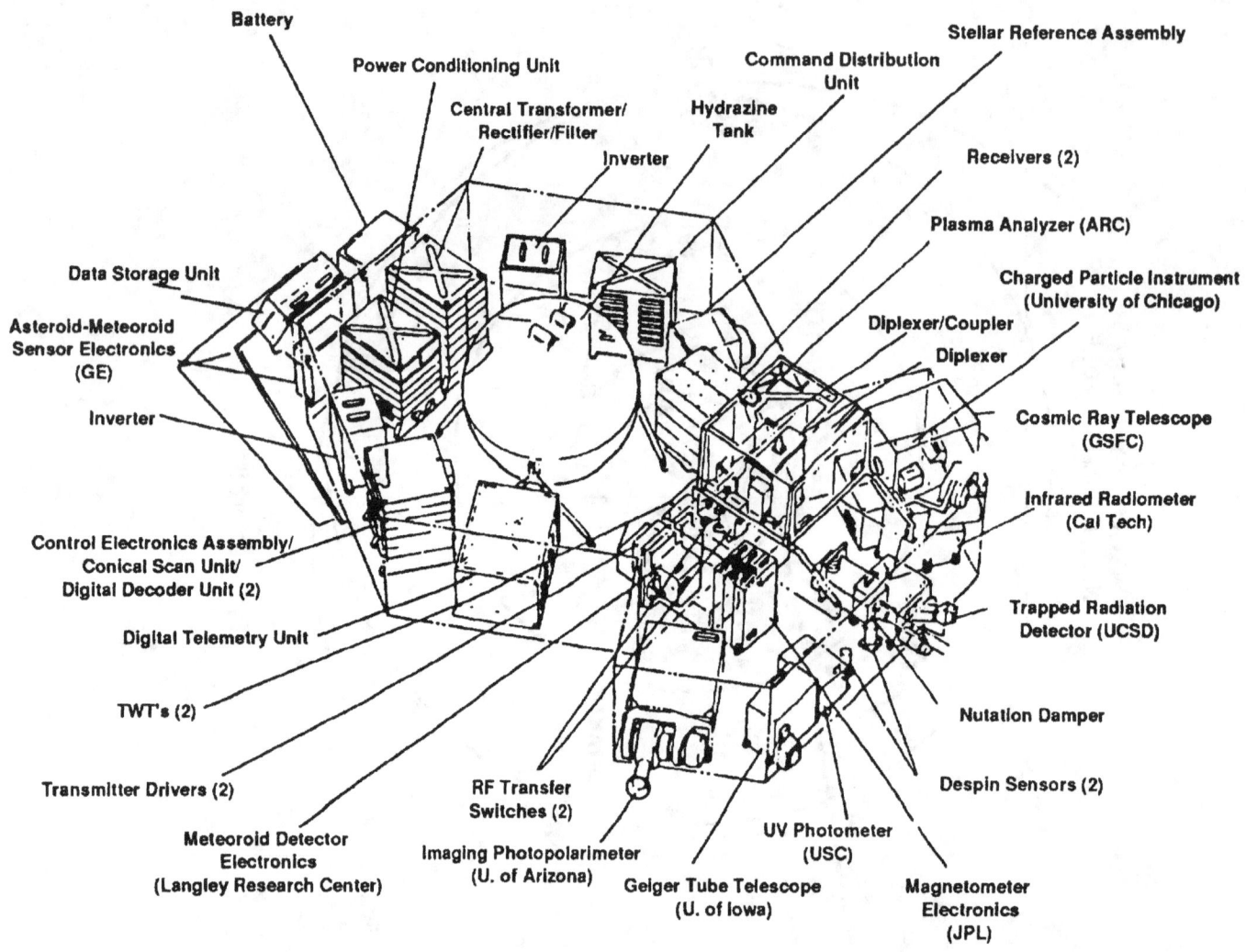

Internal arrangement of spacecraft equipment compartment

Section 3 Operational Spacecraft

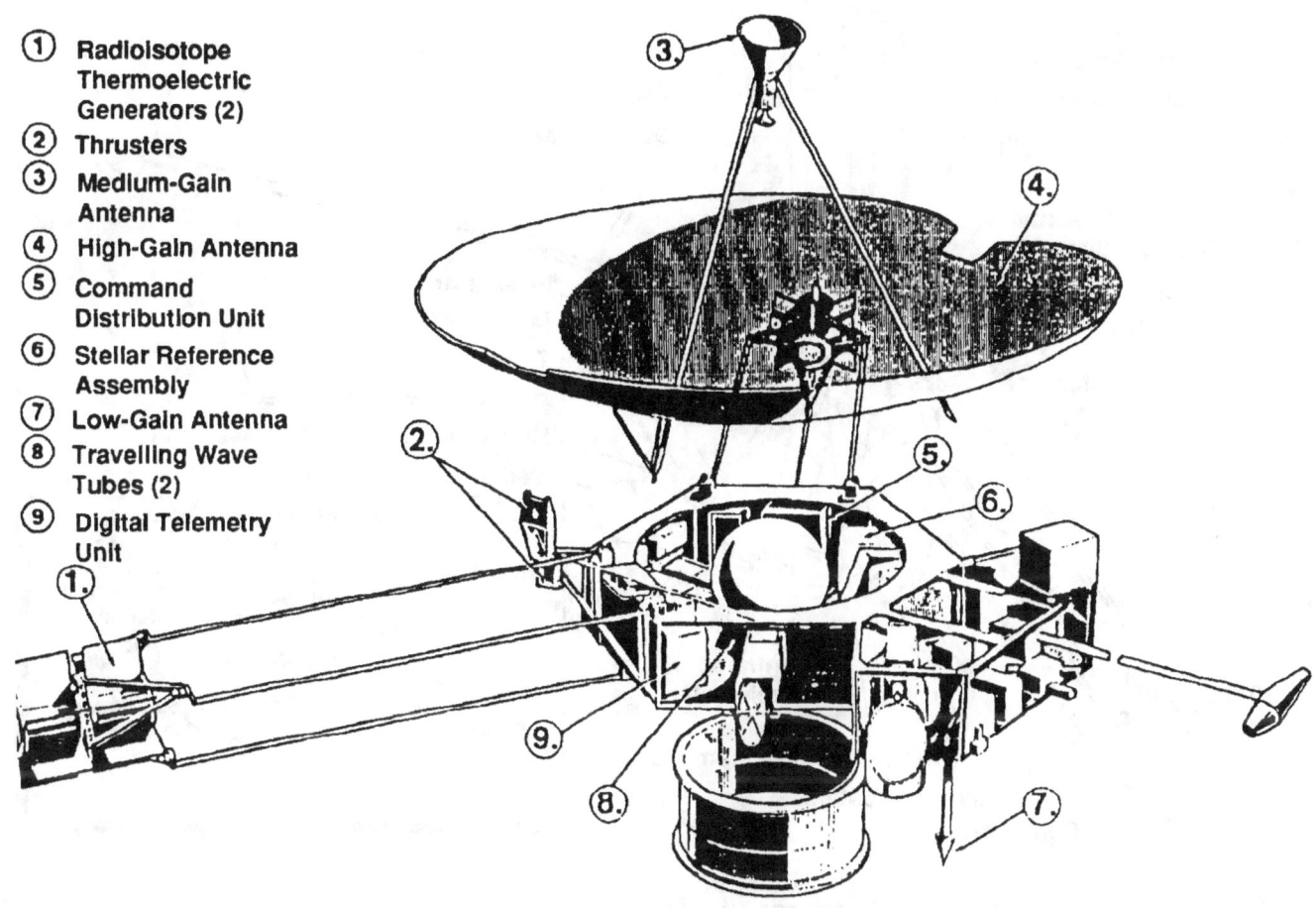

1. Radioisotope Thermoelectric Generators (2)
2. Thrusters
3. Medium-Gain Antenna
4. High-Gain Antenna
5. Command Distribution Unit
6. Stellar Reference Assembly
7. Low-Gain Antenna
8. Travelling Wave Tubes (2)
9. Digital Telemetry Unit

Internal view of major spacecraft subsystems

Section 3 Operational Spacecraft

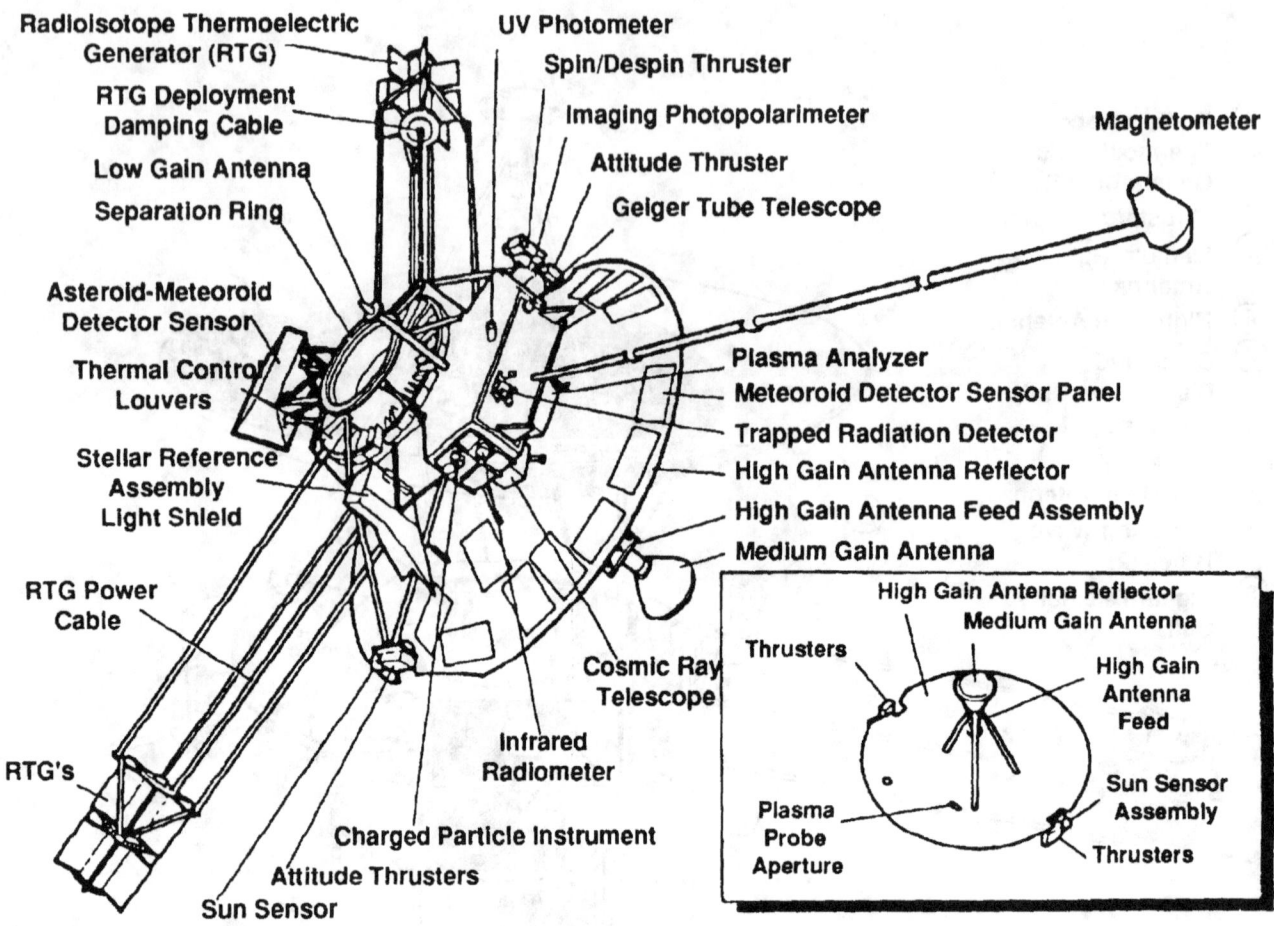

External view of the spacecraft

3.2 Interplanetary Monitoring Platform-8 (IMP-8)

Target: Near-Earth environment

Orbit: Geocentric elliptical; apogee 240,000 km, perigee 192,000 km, 11.9° Inclination

Mission Duration: Extended

Mission Class: Moderate

Mass: 401 kg

Launch Vehicle: Delta, October 1973

Theme: Global Geospace Science (GGS) Program/magnetospheric physics

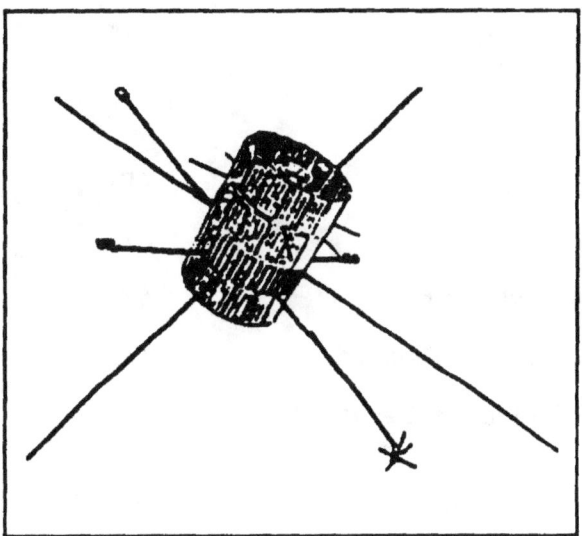

Science Objectives:

To perform detailed and near-continuous studies of the interplanetary environment for orbital periods comparable to several rotations of active solar regions.

Spacecraft:

Type: 16-sided drum, spin stabilized at 22.3 rpm

Special Features: Lacks orbit maneuver capability

Special Requirements: Spin axis normal to the ecliptic plane within a degree

Instruments:

Investigation	Experiment code	Mass (kg)	Data rate (bps)	Principal Investigator
Magnetic Fields	GNF	3.2		N.F. Ness (GSFC)
DC Electric Fields	GAF	11.5		T.L. Aggson (GSFC)
AC Electric & Magnetic Fields	IOF	12.0		D.A. Gurnett (Iowa)
Solar and Cosmic Ray Particles	GME	11.0		F.B. McDonald (GSFC)

Continued on next page

Section 3 Operational Spacecraft

Table continued

Investigation	Experiment code	Mass (kg)	Data rate (bps)	Principal Investigator
Cosmic Ray & Solar Flare Isotopes	CHE	7.4		J.A. Simpson (Chicago)
Energetic Particles	GWP	3.3		D.J. Williams (NOAA)
Charged Particles	APP	3.9		S.M. Krimigis (APL)
Electron Isotopes	CAI	8.0		E.C. Stone (JPL)
Ion and Electron	MAE	7.0		G. Gloeckler (Maryland)
Low Energy Particles	IOE	2.6		L.A. Frank (Iowa)
Solar Plasma Electrostatic Analyzer	LAP	6.3		S.J. Bame (LA SL)
Solar Plasma Faraday Cup	MAP	6.5		H.S. Bridge (MIT)

Mission Strategy:

IMP-8 was launched in October 1973 into a 12.5 day geocentric orbit with apogee and perigee near 40 Re and 30 Re respectively. The spacecraft has exceeded its planned mission life and is now operating in an open-ended extended mission in its original orbit. The orbit has been stable over the past 16 years, but the inclination varies from -55° to +55° with a period of many years.

The IMP series of spacecraft has been one of the most scientifically productive series of spacecraft ever launched. Explorer 18, which was the first of the IMP series, verified that the magnetosphere was separated from the solar wind by a standing shock wave in the solar direction. Beginning with Explorer 18, the IMPs systematically mapped Earth's radiation environment, the near-Earth solar wind, and the details of the magnetospheric collisionless shock wave. IMP spacecraft have been placed in deep interplanetary Earth orbits and IMP-E (Explorer 35) was put into orbit around the Moon, thereby providing detailed observations of the Moon's particles and fields environment. The IMPs also explored the tail of Earth's magnetosphere to better understand the flow of plasma and magnetic fields around and away from Earth. The IMP series of Explorers consisted of 10 spacecraft, all of which were exceptional scientific successes.

Enabling Technology Development: None

Section 3 Operational Spacecraft

Points of Contact:

Program Manager:	Jim Willett (202) 453-1514
Program Scientist:	Mary Mellot (202) 453-1514
NASA Center:	GSFC
Project Manager:	Paul Pashby
Project Scientist:	Joseph H. King (301) 286-7355

Section 3 Operational Spacecraft

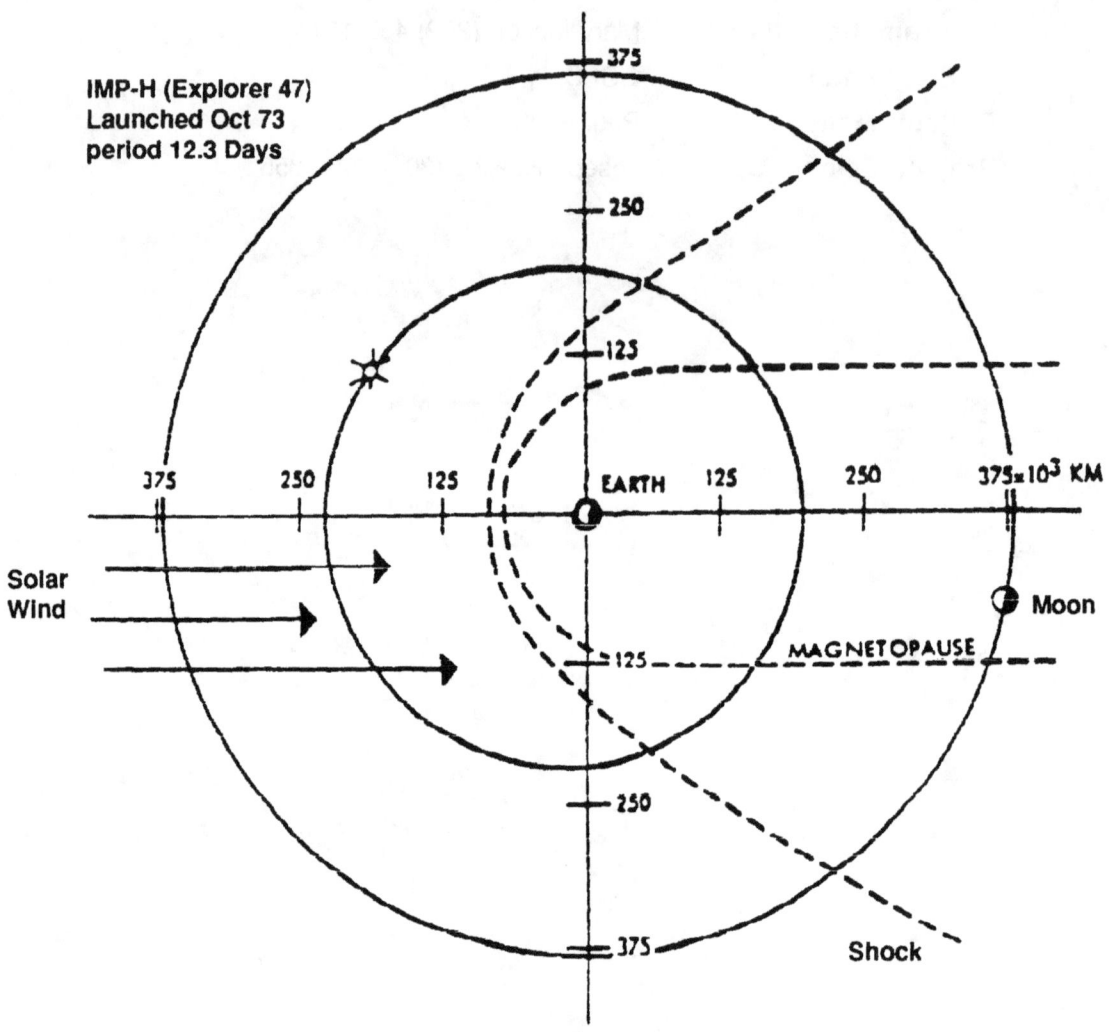

IMP-8 orbit shown with Moon's orbit,
magnetosphere bow shock, and magnetotail

Section 3 Operational Spacecraft

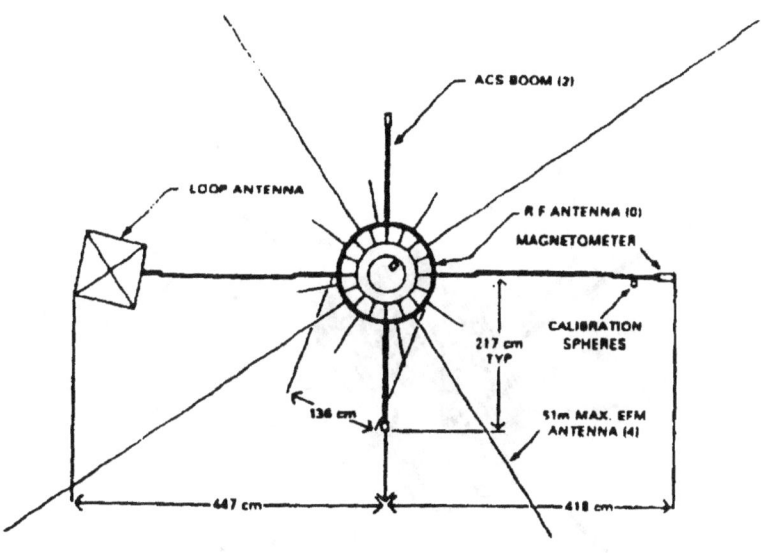

IMP appendages

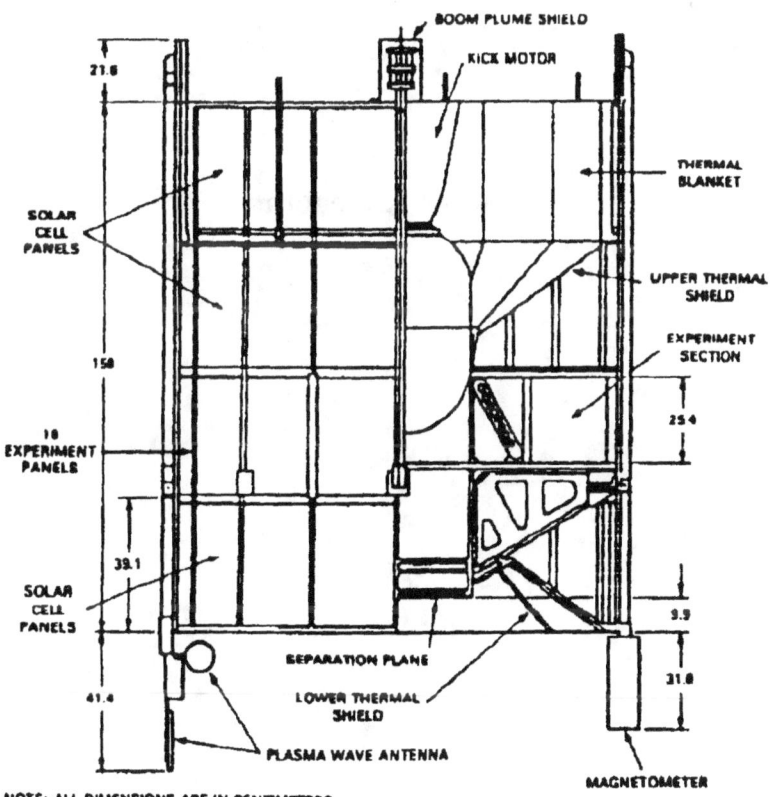

IMP structural launch configuration

Section 3 Operational Spacecraft

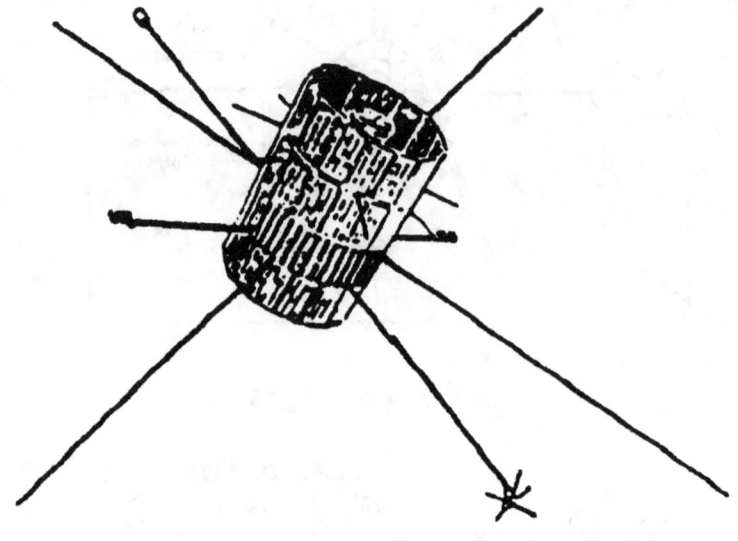

IMP-8 spacecraft

3.3 Voyagers 1 and 2

Target: Large-scale properties of the heliosphere/interstellar space

Orbit: Solar system escape trajectory

	Voyager 1	Voyager 2
Vα	3.501 AU/yr	3.386 AU/yr
Ecliptic Lat.	35.549°	-47.455°
Ecliptic Long.	260.778°	310.885°

Mission Duration: To RTG expiration

Mission Class: Moderate

Mass: 800 kg

Launch Vehicle: Titan 3E-Centaur
5 September 1977—Voyager 1
20 August 1977—Voyager 2

Theme: Cosmic and heliospheric physics

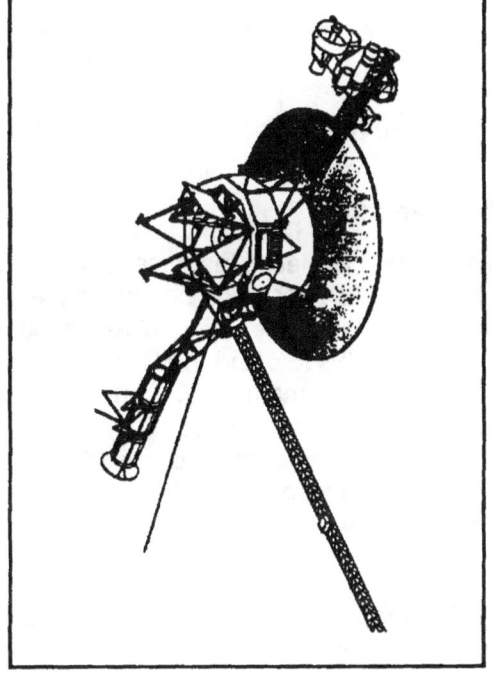

Science Objectives:

Space Physics Division will assume program management responsibility for Voyager 1 and 2 spacecraft during FY 93. The science objectives during the Voyager Interstellar Mission (VIM) are as follows:

- Characterize the solar wind with distance from the Sun.

- Observe and characterize the Sun's magnetic field reversal.

- Search for low-energy cosmic rays.

- Characterize particle acceleration mechanisms in the interplanetary medium.

- Search for evidence of interstellar hydrogen and helium and an interstellar wind .

- Locate the heliospheric/interstellar boundary.

Spacecraft:

Type: 3-axis stabilized, with mono-propellant hydrazine thrusters

Special Features: Offset pointing capability provided by gyros. Reprogrammable computers with capability for mission changes.

Special Requirements: TBD

Instruments:

There are 11 science instruments on the two Voyagers. All but four are located on the scan platform or its supporting boom. Of these four, the magnetometer uses its own boom; the Planetary Radio Astronomy (PRA) experiment shares an antenna with the Plasma Wave Subsystem (PWS); and the Radio Science Subsystem (RSS) uses the radio beams from the High Gain Antenna (HGA). Four instruments on the scan platform require accurate pointing: the Imaging Science Subsystem (ISS) wide and narrow-angle cameras; the Ultraviolet Spectrometer; and the Photopolarimeter Subsystem (PPS). The remaining three instruments on the scan platform boom—all fields and particles experiments—are the Cosmic Ray Subsystem (CRS), the Low-Energy Charged Particle (LECP) experiment, and the Plasma Subsystem (PLS).

During the interstellar cruise phase, the four fields and particles instruments will be used as follows:

- The CRS will measure the energy spectrum of particles between 1 and 500 MeV.

- The LECP will investigate the very low-energy end of the spectrum by differentiating charged particles by source, composition, energy, flux intensity, and favored direction.

- The PLS will be concerned with the collective properties of very hot ionized plasma, such as the solar wind, determining its intensity, density, pressure, and flux direction.

- The Magnetometer System will be able to detect the heliopause directly by a sudden drop in full intensity.

Mission Strategy:

Voyager 1 is currently (Jan. 1991) at a distance of 44 AU with a velocity of 16.6 Km/s, while Voyager 2 is at 34 AU with a velocity of 16.05 Km/s. Both spacecraft are exiting the solar system in the same general direction as Pioneer 11, with Voyager 1 rising steeply above the ecliptic plane at 35.5° and Voyager 2 descending even more steeply below the ecliptic at -47.5°.

The two Voyagers, along with the two Pioneers, are in an excellent position to provide some early answers to the heliopause/interstellar medium investigation—all four are headed out of the solar "bubble," but in four different directions. Communications will be maintained as long as the spacecraft continue to function. It is expected that the Pioneer will enable prediction of the boundary between the solar wind and the interstellar medium, allowing measurements to be made of the interstellar fields and particles unmodulated by the solar plasma.

The Sun sensor may exceed its design limitations at about 8 AU. This may happen at about 80 AU in the year 2001 for Voyager 1 and 2006 for Voyager 2, although there is a good chance the sensor will continue to function well beyond 80 AU. Thereafter, declining hydrazine reserves and/or minimal power requirements of 230 watts would be reached in about 2015, when Voyager 1 and 2 would be at heliocentric distances of 130 and 110 AU respectively.

Enabling Technology Development: None

Points of Contact:

Program Manager:	Jim Willett (202) 453-1514
Program Scientist:	Vernon Jones (202) 453-1514
NASA Center:	JPL
Project Manager:	George Textor
Project Scientist:	Ed Stone

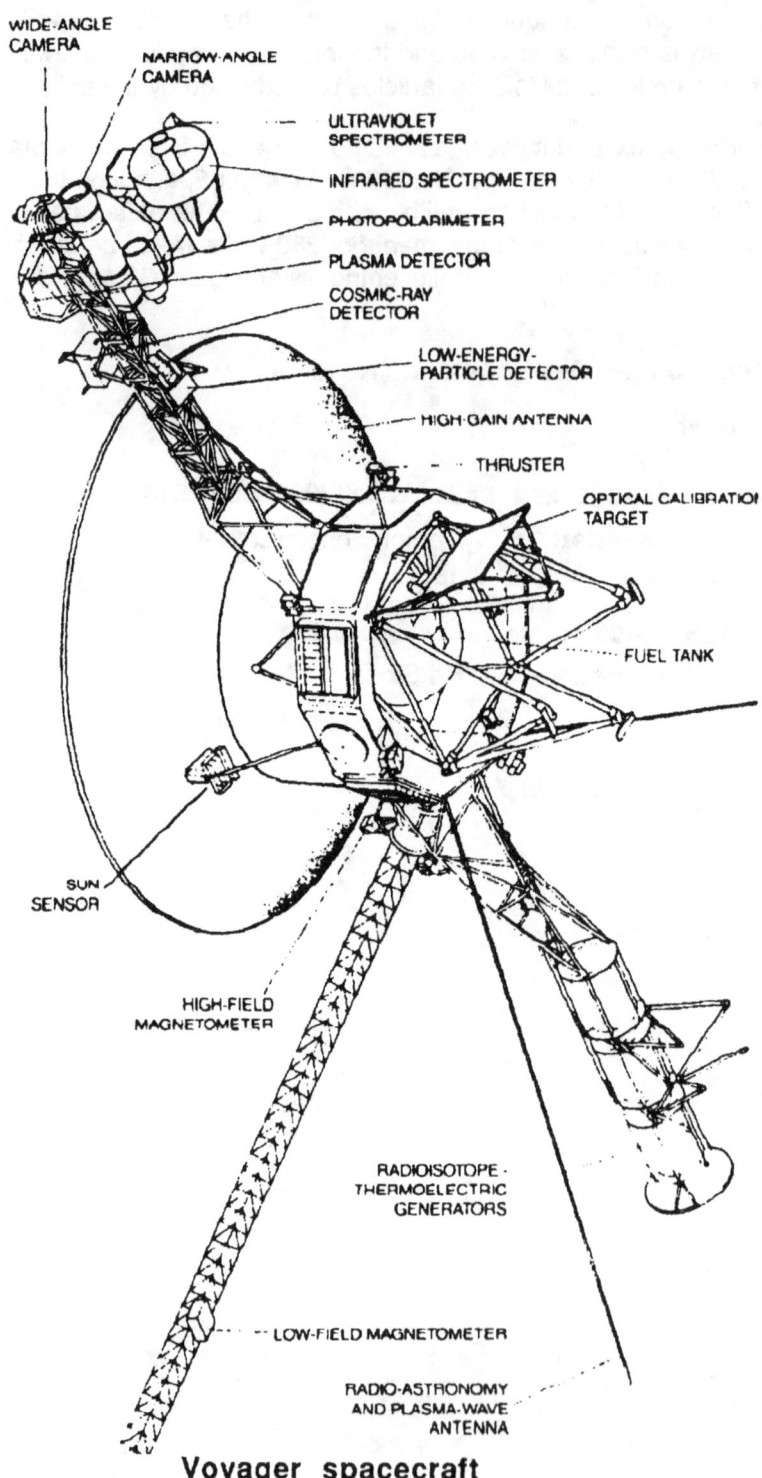

Voyager spacecraft

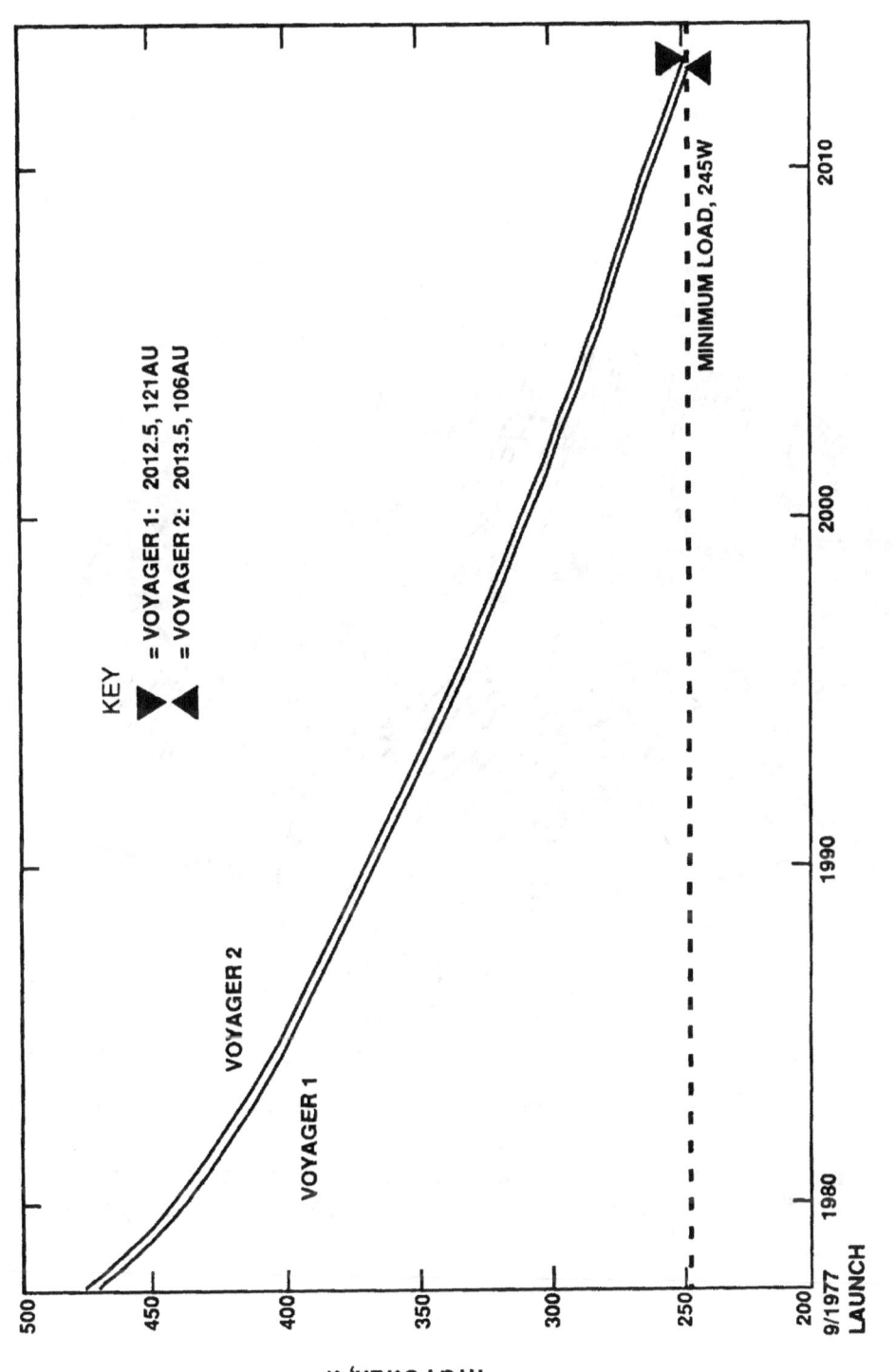

Expected profiles of RTG power output aboard Voyager spacecraft

Section 3 Operational Spacecraft

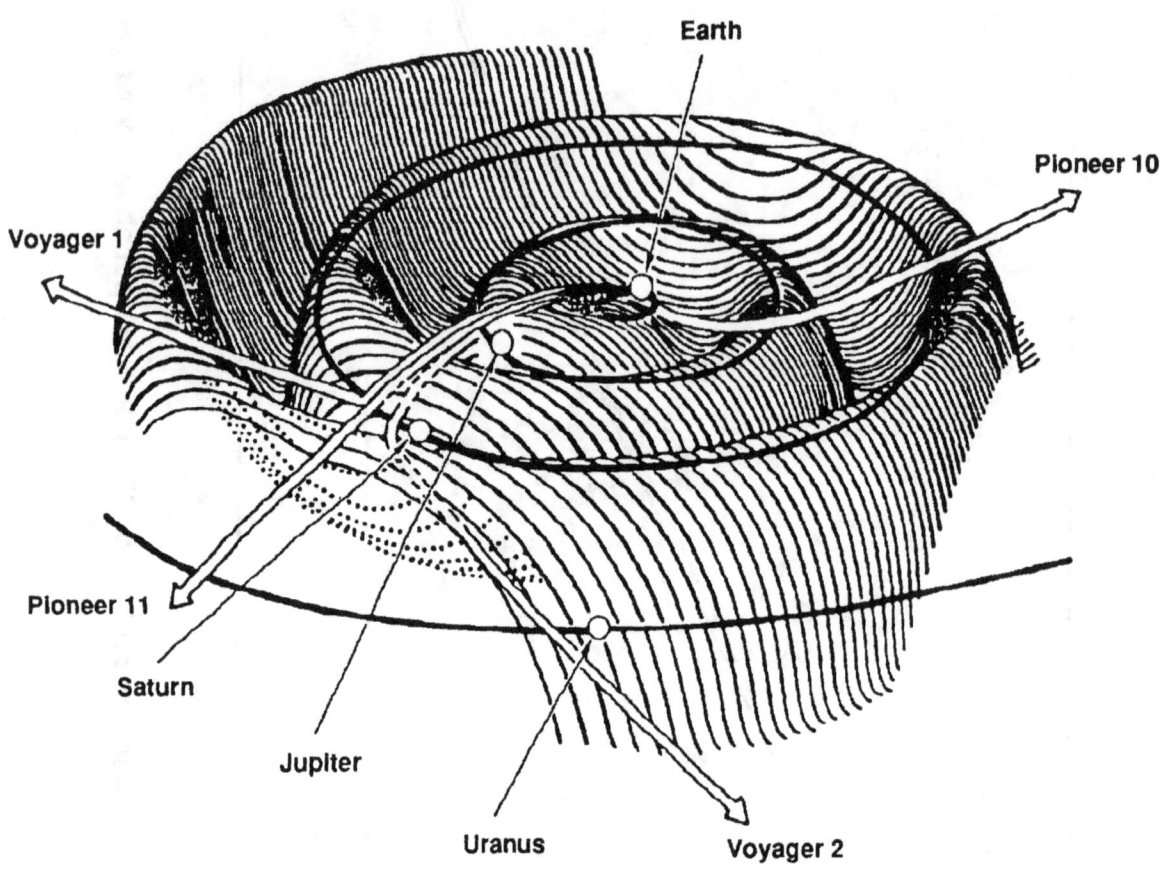

Spiral solar current sheet shown out to 20 AU: spacecraft observe magnetic field polarity reversal at current sheet crossing

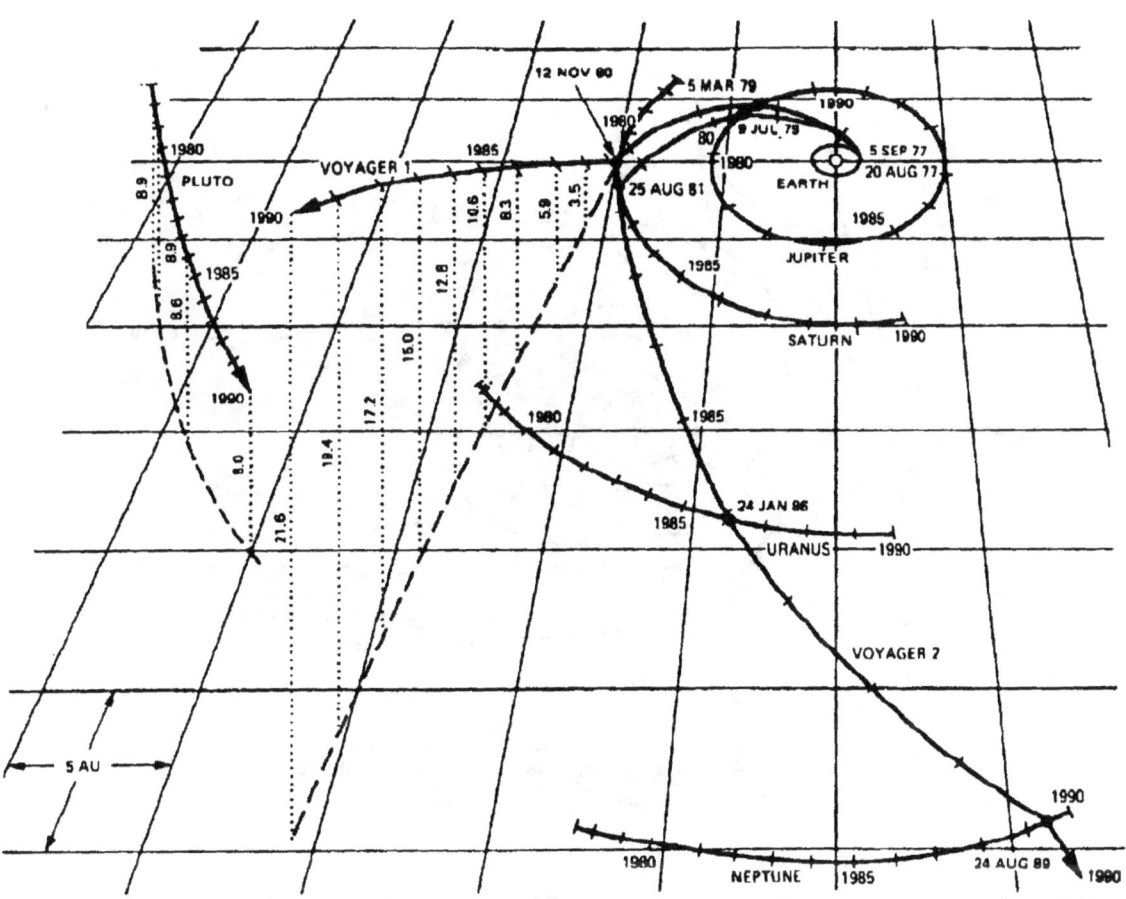

Voyager orbits

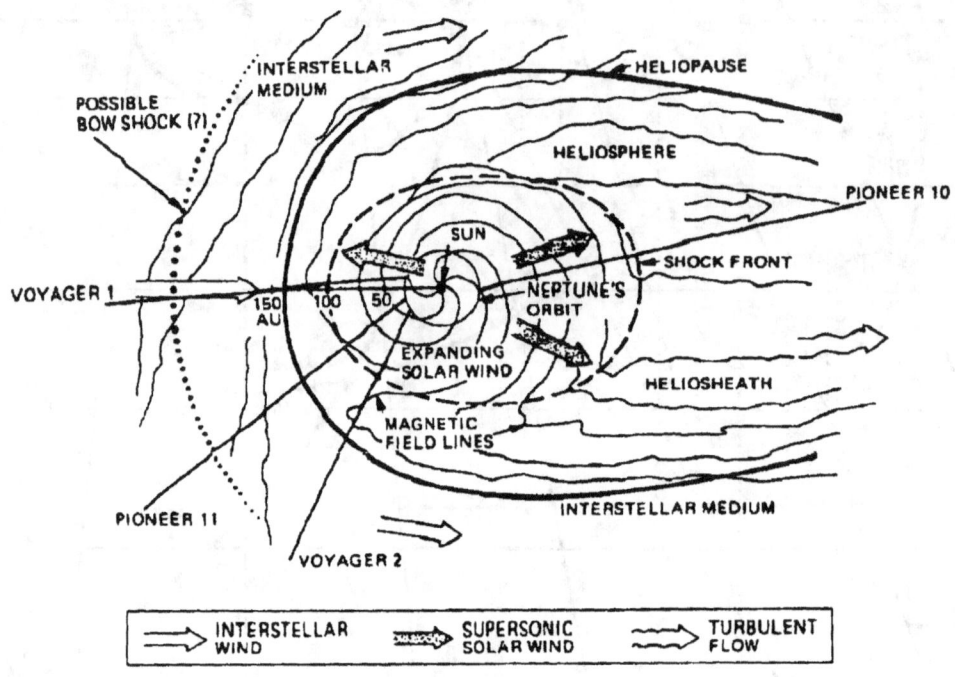

Hypothetical heliosphere model, projected into solar equator plane (after E.J. Smith) and showing trajectories of solar system escaping spacecraft

3.4 International Cometary Explorer (ICE)

Target: Heliosphere and cosmic rays at ~1 AU

Orbit:

The ICE heliocentric orbit is very much like that of Earth, with an eccentricity of roughly 0.05 and an inclination to the ecliptic of approximately 0.06°. It has a perihelion radius of about 0.926 astronomical units (AU) and an aphelion radius of about 1.033 AU. The orbital period is approximately 354 days, which means that ICE orbits the Sun approximately one and one-thirtieth times in a year, thereby setting up the August 2014 Earth-return opportunity. As of July 20, 1989, ICE was 1.179 AU distant from the Earth.

Mission Duration: Open-ended

Mission Class: Moderate

Mass: 417.9 kg (current) 478 kg (launch)

Launch Vehicle: Delta 2914, 12 August 1978

Theme: Cosmic and heliospheric physics

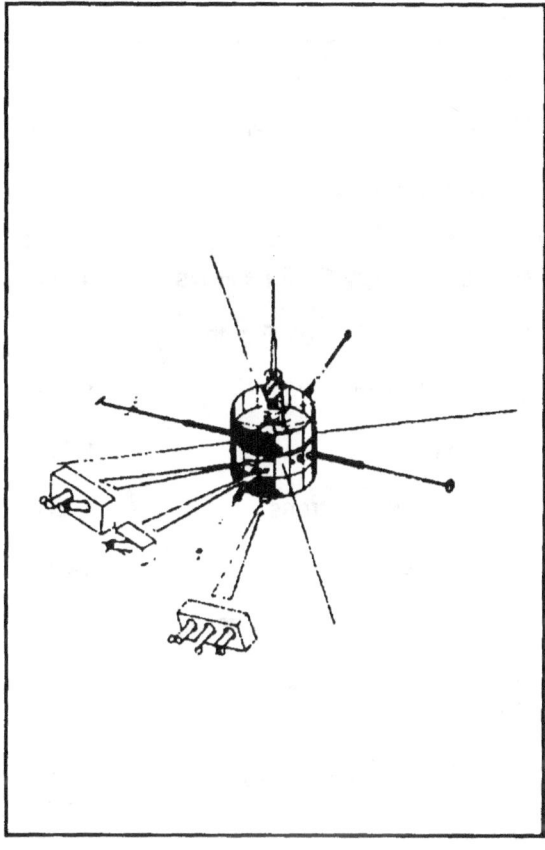

Science Objectives:

The science objectives of ICE are to investigate the properties of magnetic fields, plasma, flare energetic particles, IP shocks, Type II & III radio bursts, and galactic cosmic rays.

Spacecraft:

    Type: Large Explorer, spin stabilized (20 rpm)
    Special Features: +Z-axis tower, Y-axis inertia booms
    Special Requirements: Spin axis to be within 1° of North ecliptic pole

Instruments:

Investigation	Expt code	Mass (kg)	Data rate (bps)	Principal Investigator
Solar Wind Plasma	BAH			S. Bame (Los Alamos)
Helium Vector Management	SMH			E. Smith (JPL)
Energetic Protons	DFH			R. Hynds (Imp. College, London)
Medium Energy Cosmic Rays	TYH			T. Von Rosenvinge (GSFC)
X-Rays, Low Energy Electrons	ANH			K. Anderson (UCB)
Plasma Waves	SCH			F. Scarf (TRW)
Radio Waves	SBH			J. Steinberg (Meuden)
Cosmic Ray Electrons	MEH			P. Meyer (Univ. of Chicago)
Plasma Composition	OGH			K. Ogilve (GSFC)
High Energy Cosmic Rays	STH			E. Stone (CIT)
High Energy Cosmic Rays (Draft chamber failed)	HKH			M. Wiedenbeck (Univ. of Chicago)
Low Energy Cosmic Rays	HOH			D. Hovestadt (MPI)

Mission Strategy:

The ISEE-3 spacecraft was launched on August 12, 1978 and injected into a 100-day long transfer trajectory out to the vicinity of the Sun-Earth L_1 libration point, where it was subsequently inserted into a large-amplitude halo orbit around the L_1 point. ISEE-3 was maintained in the 6-month period halo orbit until June 1982 when a deorbit maneuver placed it on a transfer trajectory back to the vicinity of the Earth to begin a new phase called the Extended Mission. The Extended Mission phase was a multiple double-lunar-swingby (DSL) trajectory with outer loops designed to traverse the geomagnetic tail and apogee locations deep in the tail (maximum distance reached was 236.6 Earth radii). The Extended Mission phase lasted until December 22, 1983, when final, controlled swingby of the Moon gave ISEE-3 the orbital energy boost needed to leave the Earth-Moon system and go into a pre-designed orbit around the Sun. This new heliocentric trajectory was designed for a flyby of the Comet Giacobini-Zinner (G-Z) as the latter made its descending-node passage of the ecliptic plane on September 11, 1985. The ISEE-3 spacecraft, renamed the International Cometary Explorer (ICE) following the final lunar swingby, had a highly successful encounter with Comet G-Z (penetrating both the coma and tail of the comet with no apparent damage and transmitting data the entire time). Currently, the ICE spacecraft is still operational, and its trajectory will bring it back to the Earth-Moon system in August 2014, when an energy-robbing leading-edge swingby of the Moon and subsequent recapture into geocentric orbit should be possible.

The spacecraft will return to a point that is 1 AU from the Earth in the year 2008. At that time, it is planned to re-target ICE for a Lunar gravity-orbit maneuver in August 2013 that will put the spacecraft into high-energy Earth orbit.

Enabling Technology Development: None

Points of Contact:

Program Manager:	Jim Willett
Program Scientist:	Mary Mellot
NASA Center:	GSFC
Project Manager:	Paul Pashby
Project Scientist:	Keith Ogilvie

Section 3 Operational Spacecraft

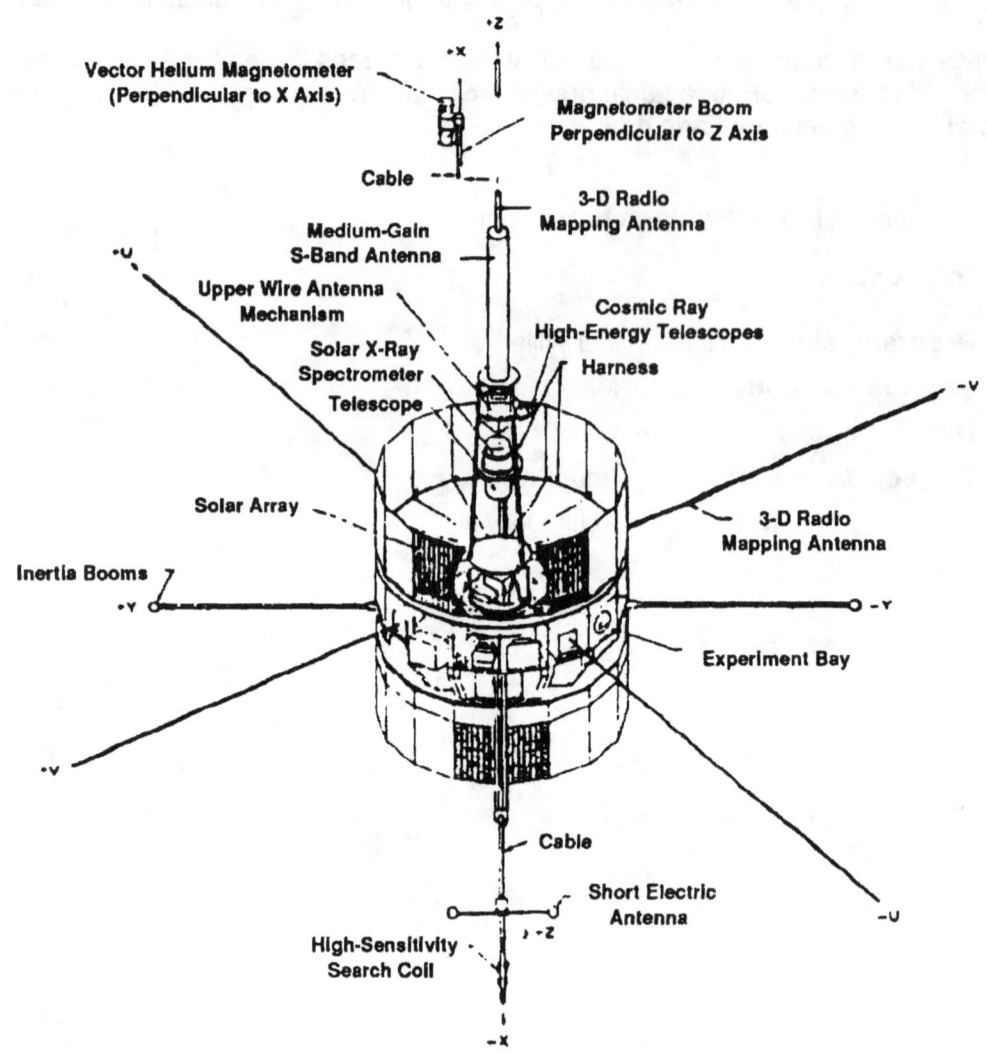

ICE-configuration showing experiments, communications tower, and body axis system

Section 3 Operational Spacecraft

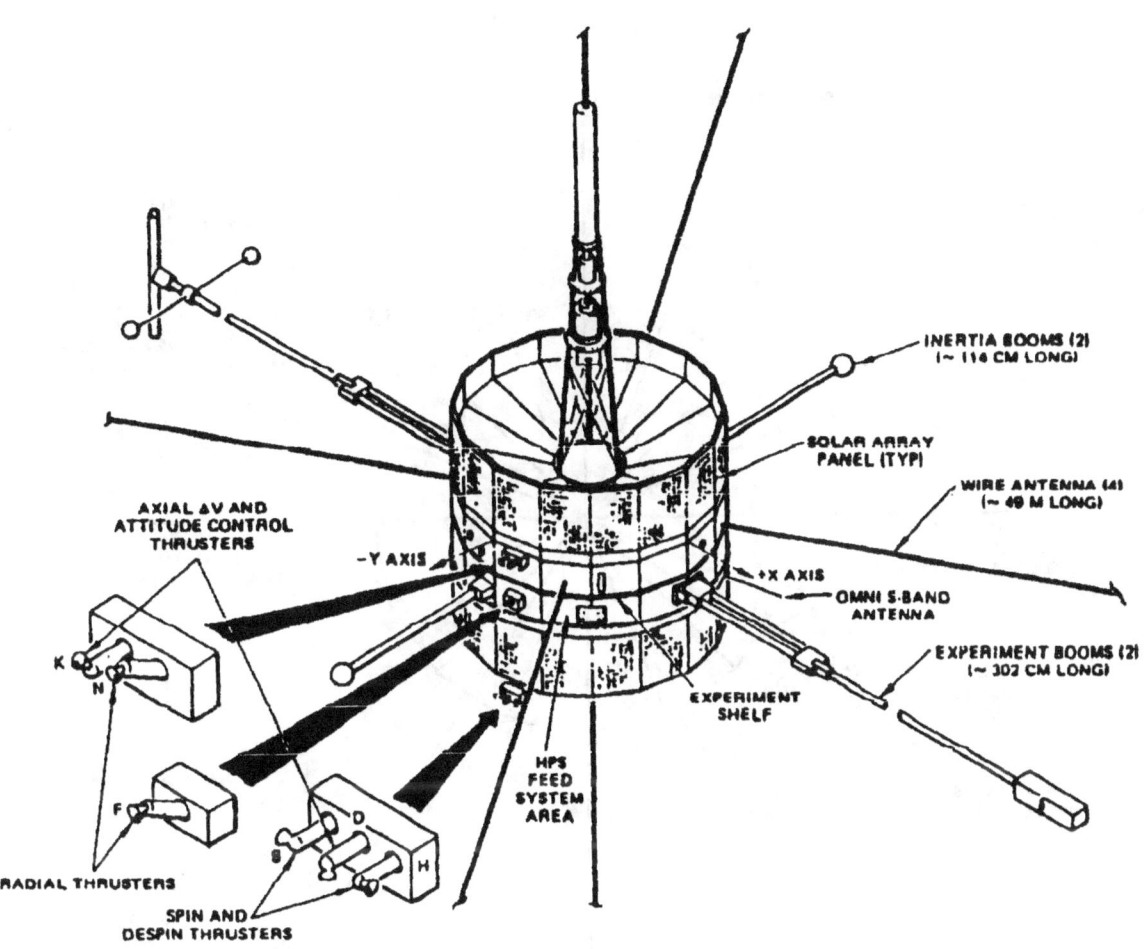

**ICE attitude and orbital control system pictorial
(an identical system of thrusters is located on the facet
opposite to the one shown)**

Section 3 Operational Spacecraft

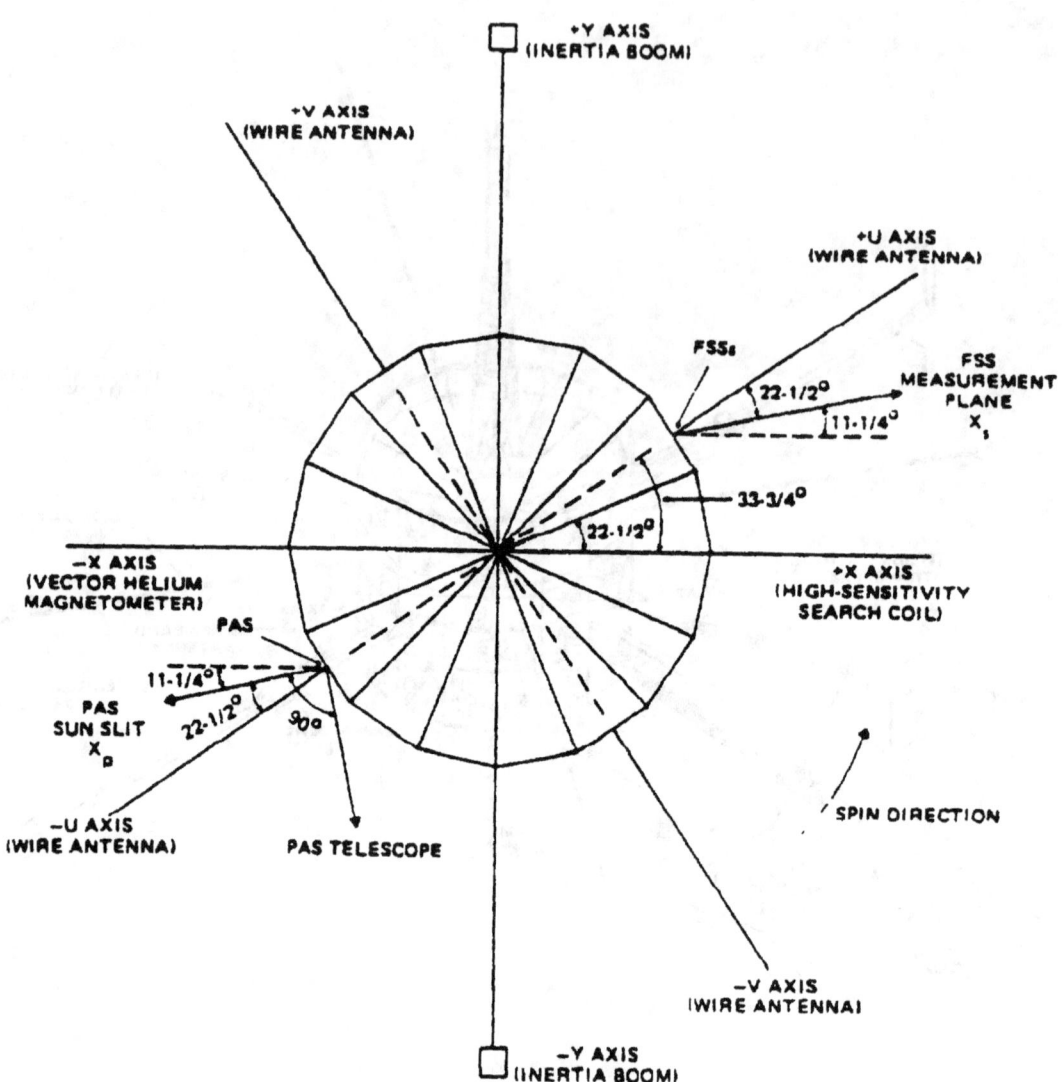

Top view of ICE with attitude sensor data

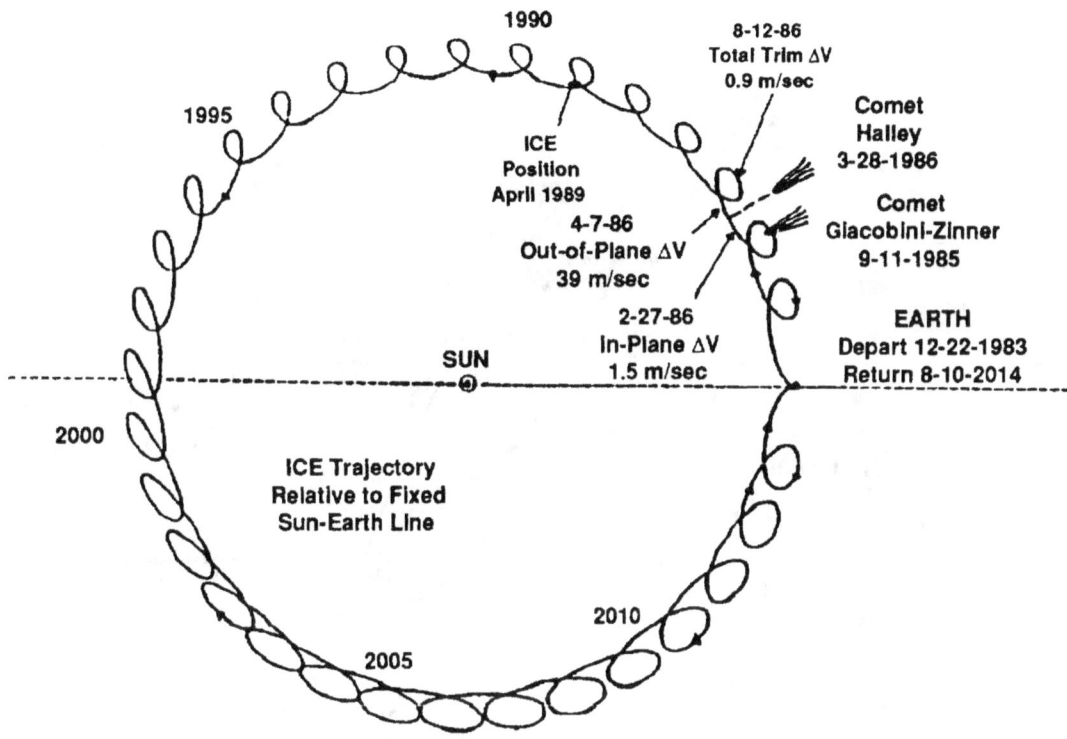

The ICE Earth-return trajectory (1983 to 2014) plotted relative to a fixed Sun-Earth line. Time interval between departure and return lunar swingbys is 11,189 days, or 30.634 years.

Start and End Dates/Durations:

Phase	Dates	Duration
Transfer trajectory to halo insertion	8–12–78 to 11–20–78	100 days
Halo orbit phase	11–20–78 to 9–1–82	3 years, 9 months
Extended mission	9–1–82 to 12–22–83	1 year, 3.7 months
Comet intercept mission	12–22–83 to 9–11–85	1 year, 8.7 months
Heliocentric cruise	9–11–85 to present	Open ended

Section 3 Operational Spacecraft

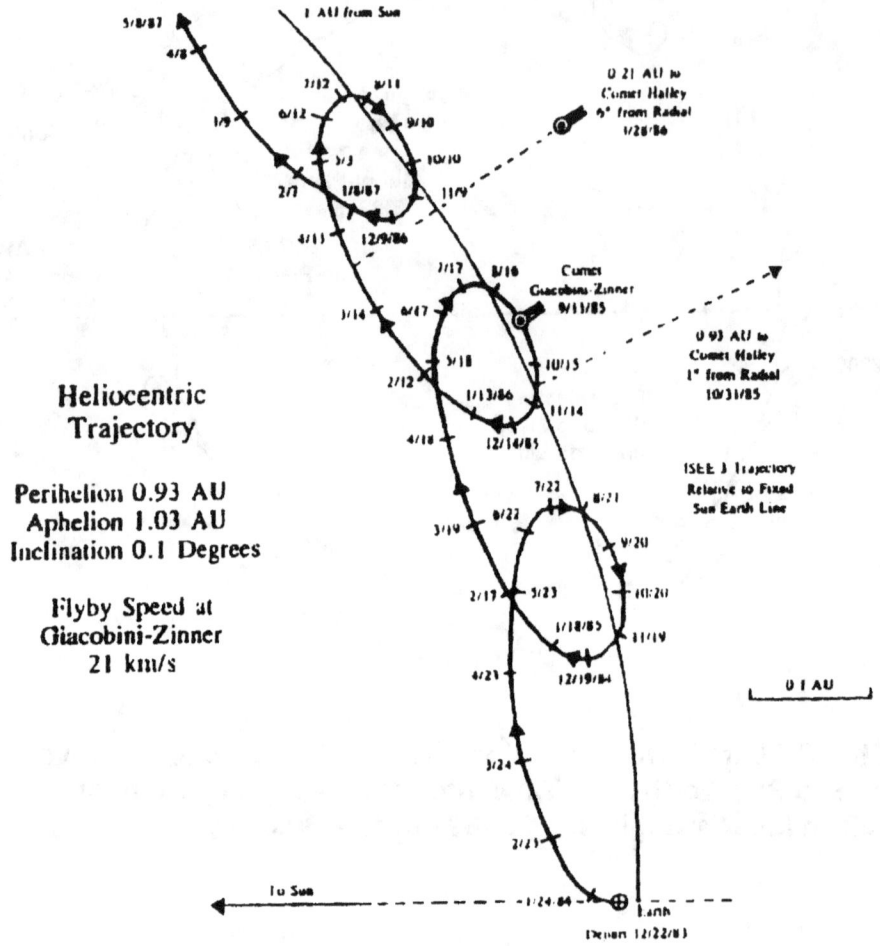

ICE heliocentric trajectory

3.5 Dynamics Explorer-1 (DE-1)

Target: Magnetosphere/ionosphere coupling

Orbit: Polar orbit 23170 x 570 km initial orbit

Mission Duration: Continues

Mission Class: Small Explorer

Mass: 423 kg (at launch)
 105 kg (payload)

Launch Vehicle: Delta 3913
 (August 1981)

Theme: Magnetospheric physics

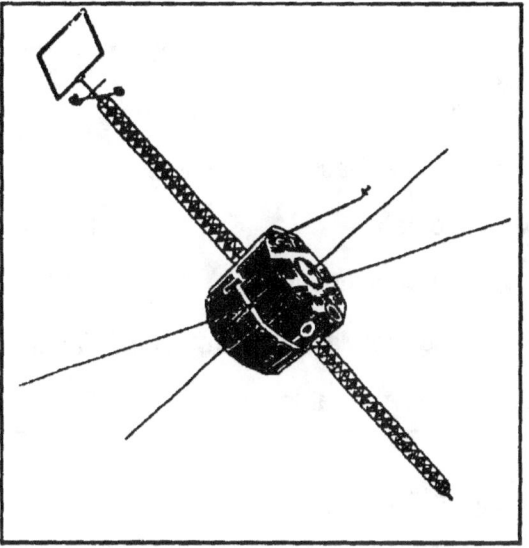

Science Objectives:

The objective of the Dynamics Explorer (DE) program is to investigate the strong interactive processes coupling the hot, tenuous, convecting plasmas of the magnetosphere and the cooler, denser plasmas and gases co-rotating in the Earth's ionosphere, upper atmosphere, and plasmasphere.

Spacecraft:

Type: Explorer, spin stabilized at 10 rpm
Special Features: TBD
Special Requirements: TBD

Instruments:

Investigation	Expt code	Mass (Kg)	Data rate (bps)	Principal Investigator
Energetic Ion Composition Spectrometer	EICS			Shelley (Lockheed)
High Altitude Plasma Instrument	HAPI			Burch (SWRI)
Magnetometer	MAG			Seguira (MSFC)
Plasma Wave Instrument	PWI			
Retarding Ion Mass Spectrometer	RIMS			Chappel (MSFC)
Spin-Scan Auroral Imager	SAI			Frank (Univ. Iowa)

Mission Strategy:

The mission originally consisted of two spacecraft placed in polar, coplanar orbits: DE-1 at a perigee of approximately 570 km and apogee of approximately 23,170 km, and DE-2 at a perigee of approximately 309 km and an apogee of approximately 1012 km. DE-1 and DE-2 were dual-launched in August 1981. At this time DE-1 is still operational, but DE-2 decayed in February 1983.

Enabling Technology Development: None

Points Of Contact:

Program Manager:	Jim Willett
Program Scientist:	Mary Mellot
NASA Center:	GSFC
Project Manager:	Paul Pashby
Project Scientist:	Bob Hoffman

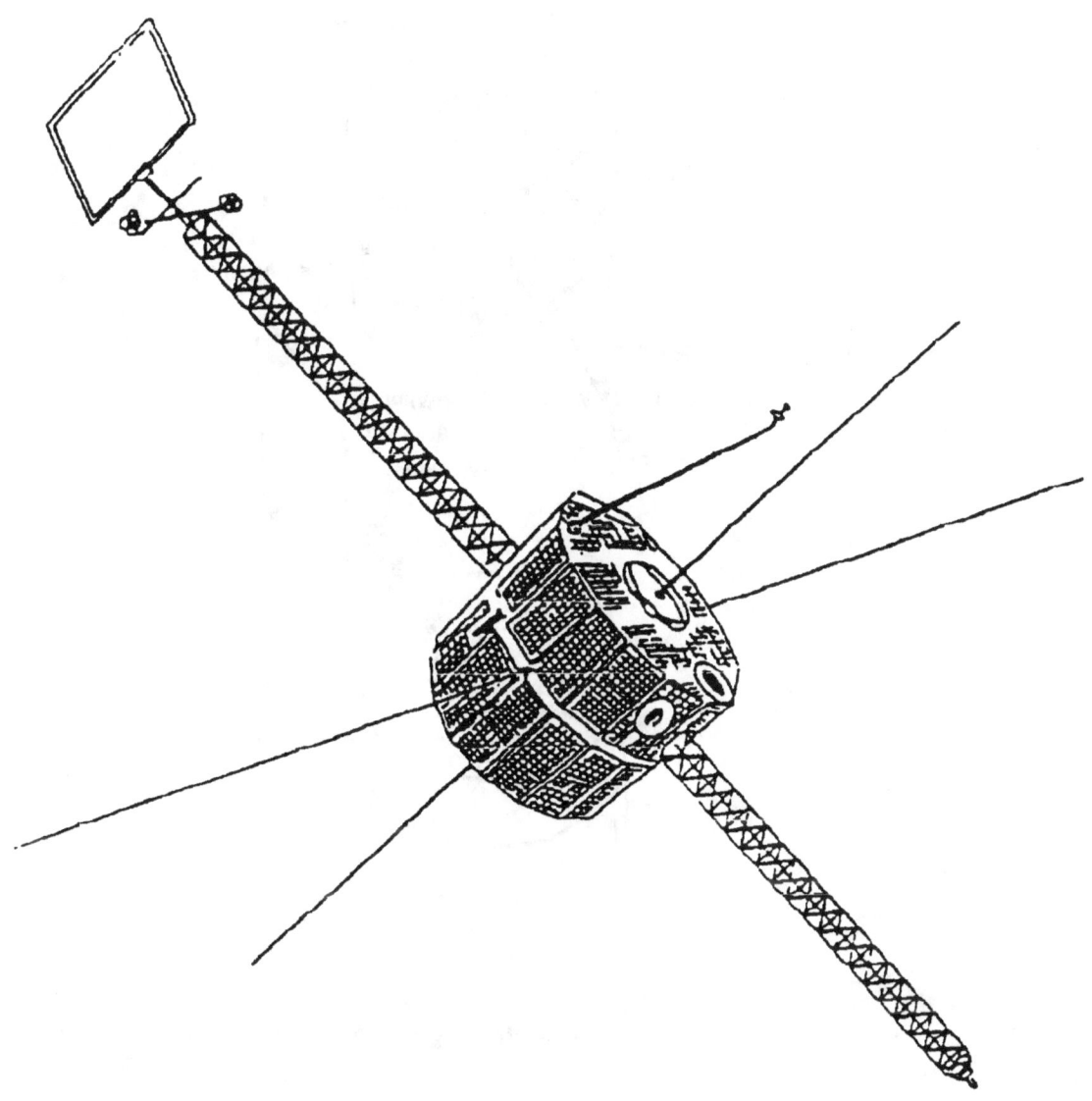

Dynamic Explorer-1

Section 3 Operational Spacecraft

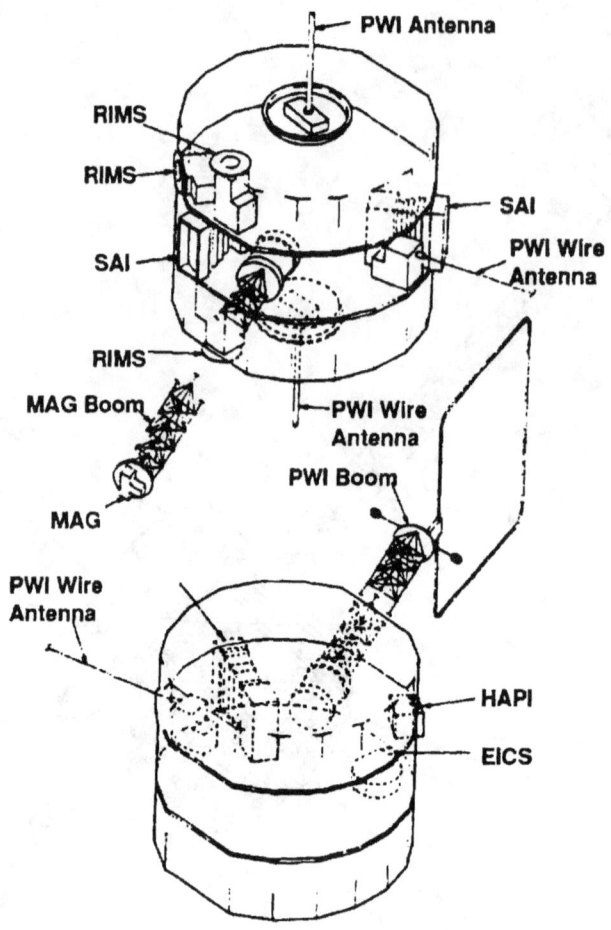

Dynamics Explorer

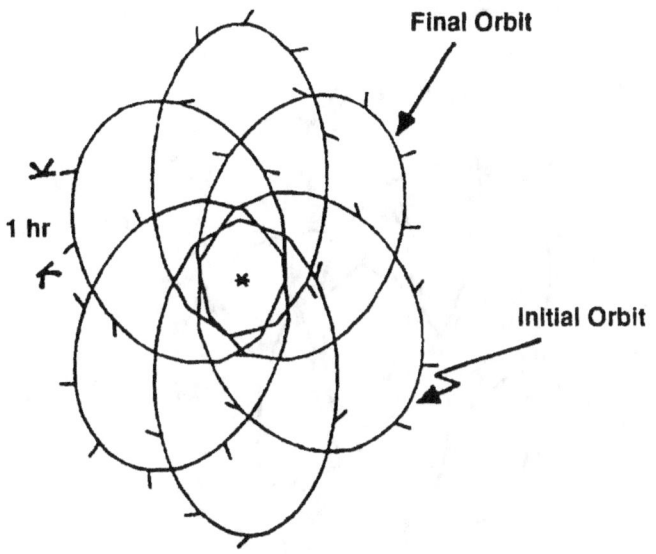

Orbit shown at 6-month intervals

	Initial orbit	Final orbit
Epoch	08/04/85	02/04/88
Ω	159°	155°
W	155°	210°
I	90°	89°

DE-1 orbit: 4.0 to 6.5 years (1.08 x 4.66 R_e fixed orbit)

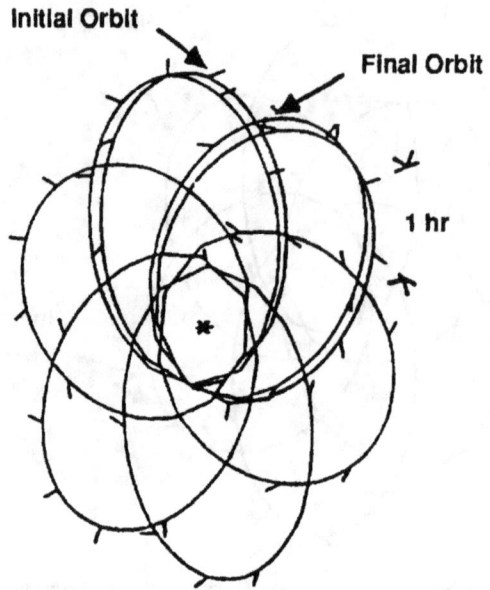

Shown at 6-month intervals

	Initial orbit	Final orbit
Epoch	08/03/81	02/03/85
Ω	161°	160°
W	282°	216°
I	90°	90°

First 3.5 years of DE-1 orbit (1.08 x 4.66 R_e fixed orbit)

3.6 Combined Release and Radiation Effects Satellite (CRRES)

Target: Ionosphere and magnetosphere

Orbit: Elliptical, 348 km x 33,582 km, at 18.15° inclination, orbit period 10.56 hrs

Mission Duration: 6 years (extended)
3 years (nominal)

Mission Class: Moderate

Mass: 1780 kg

Launch Vehicle: Atlas/Centaur
(25 July 1990)

Theme: Ionospheric physics, magnetospheric physics

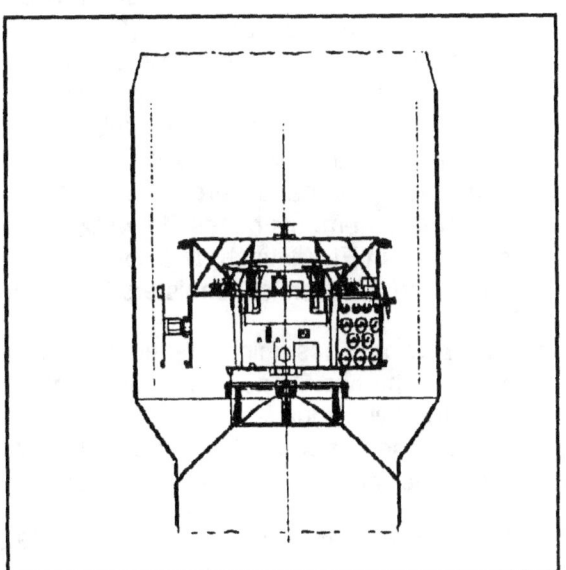

Science Objectives:

The CRRES program is a joint NASA/DoD undertaking. Management of the DoD portion is performed by the Space Test and Transportation Program Office of the United States Air Force (USAF) Space Division.

CRRES carries a complement of chemical release canisters to be released at certain times over ground observation sites and diagnostic facilities. These releases form large clouds that will interact with the ionospheric and magnetospheric plasma and the Earth's magnetic field. These interactions will be studied with optical, radar, and plasma wave and particle instruments from the ground, from aircraft, and the CRRES spacecraft.

Those controlled experiments which are performed near perigee will further the understanding of the interaction of plasmas with magnetic fields, the coupling of the upper atmosphere with the ionosphere, the structure and chemistry of the ionosphere, and the structure of low-altitude electric fields. Those which are performed near apogee in the Earth's magnetosphere will study the formation of diamagnetic cavities, coupling between the magnetosphere and ionosphere, and the effects of artificial plasma injections upon the stability of the trapped particles in the radiation belts.

Spacecraft:

Type: Ball Aerospace special design, spin stabilized at 2.2 rpm
Special Features: TBD

Section 3 Operational Spacecraft

Special Requirements:

The 3,842 pound CRRES spacecraft is configured with eight compartments for holding science electronics, chemical canisters (Ba, Li, Ca, and Sr), an electrical power subsystem (EPS); a hydrazine reaction control subsystem (RCS); an attitude determination and control subsystem (ADCS); a telemetry, tracking, and control subsystem (TT&C); a magnetometer boom assembly; and a hoop antenna boom assembly. Also, TT&C subsystem antennas, science (wire and boom) antennas, science experiments and electronics, a nutation damper subsystem, and two solar array (SA) panels are located on the lower and upper decks of the spacecraft. To meet weight requirements necessitated by the change from a Space Transportation System (STS) launch to an ELV launch, two compartments that previously held two chemical modules containing 24 chemical canisters have been removed.

The EPS provides uninterrupted +28 dc power to the spacecraft power distribution system (PDS). The two SA panels mounted on the top deck provide relatively constant power and keep the two redundant nickel-cadmium batteries charged to provide spacecraft power during Sun eclipse periods. The EPS is a highly redundant system with the solar arrays, batteries, and power control unit each fully redundant.

The command system provides up to 448 discrete commands, 32 serial digital commands, and 256 stored commands with about one second resolution and approximately 12 hours maximum delay. Data can be stored on redundant tape recorders at 16 kbps with a playback data rate of 256 kbps.

The ADCS subsystem, in conjunction with the RCS, provides for change in velocity and precession maneuvers and spacecraft spin-rate control. Precession maneuvers use the Sun and horizon sensors as a reference. Spin rate maneuvers use the Sun sensors as a reference. The horizon and Sun sensors also provide data for Sun acquisition. The attitude control electronics provides conditioned Sun sensor, horizon sensor, and magnetometer data to the telemetry data stream.

The on-board scientific experiments consist of 46 electronic boxes, 10 booms, and 24 chemical canisters. There are 16 Principal Investigators responsible for the 16 separate chemical releases (24 chemical canisters) planned at various times during the CRRES mission.

Instruments:

Investigation	Expt code	Mass (kg)	Data rate (bps)	Principal Investigator
Chemical Releases	NASA	890.0		16 PIs from 13 Institutions
Low Altitude Satellite Studies of Ionosphere Irregularities	NRL–701	28.5	13616	P. Rodriguez (NRL)
Space Radiation Experiment	AFGL–701	168.6	11027	E.G. Mullen (AFGL)
Energetic Particles and Ion Composition	ONR–307	34.4	2674	R. Vondrak (Lockheed)
Solar Flares II	ONR–604	15.0	750	J. Simpson (Univ. of Chicago)
High Efficiency Solar Panel (HESP)	AFAPL–801	5.2	63	T. Trumble (Wright Patterson AFB)

Mission Strategy:

The CRRES spacecraft was launched from the Cape Canaveral Air Force Station (Space Launch Complex 36B) by an Atlas I launch vehicle on July 25, 1990. The spacecraft was inserted into a nominal orbit of 350 x 35,786 km and near 18° inclination. Following orbit insertion, the Centaur maneuvered to the CRRES' required separation attitude, such that the spacecraft (+)Z-axis was parallel to the ecliptic plane and pointed 12° ahead of the Sun's apparent motion. The Centaur then initiated the required spin-up (2.2 + 0.2 rpm) and issued the separation command. After providing sufficient clearance from the separated spacecraft, the Centaur performed the post-separation maneuvers to avoid collision and contamination.

The spacecraft initialization phase was an approximately 30-day period after separation from the Centaur, during which the spacecraft subsystems and experiment payloads were configured and checked out in preparation for normal on-orbit operations. Vehicle checkout and initialization were accomplished within the first 72 hours of the mission. Some critical DoD science instruments were initialized as soon as possible, following separation (with a goal of within the first 24 hours). Initialization of the remaining instruments and instrument boom deployment did not begin until all subsystems had been initialized and checked out. After final initialization and check-out of the CRRES spacecraft and its instruments, normal on-orbit operations began.

Chemical releases are permitted at any time after spacecraft subsystem initialization and will normally occur after instrument initialization. The CRRES/Geosynchronous Transfer Orbit (GTO) NASA science experiments consist of three chemical release campaigns.

The first chemical release campaign was the South Pacific Critical Velocity Releases which were done during the Moon-down period approximately two months after launch when perigee was over the South Pacific at dusk local time. The CRRES/GTO High-Altitude Releases will be conducted most probably in the Moon-down period in January 1991 when apogee is in the geotail for the first time. The final campaign will be the CRRES/GTO Caribbean Releases to be conducted in the summer of 1991 when perigee is over the Caribbean at dawn.

Enabling Technology Development: None

Points Of Contact:

Program Manager:	Rick Howard/SS
Program Scientist:	Dave Evans
NASA Center:	MSFC
Project Manager:	John Store
Project Scientist:	Dave Reasoner

Section 3 Operational Spacecraft

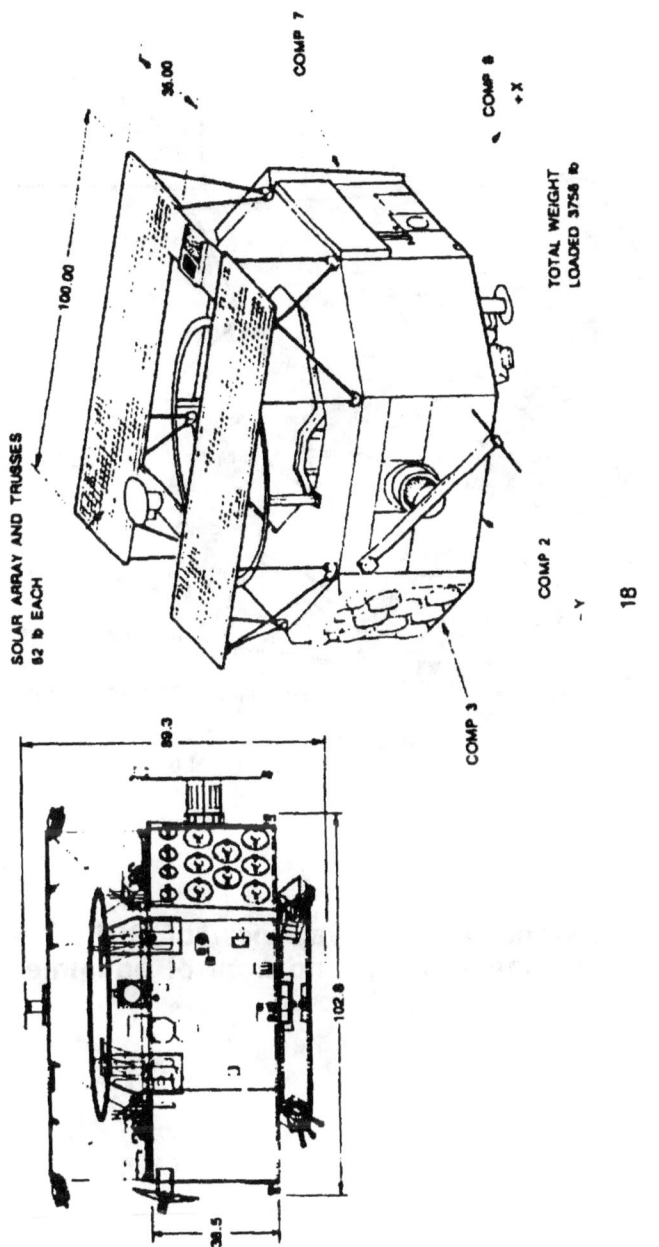

CRRES spacecraft arrangement

Section 3 Operational Spacecraft

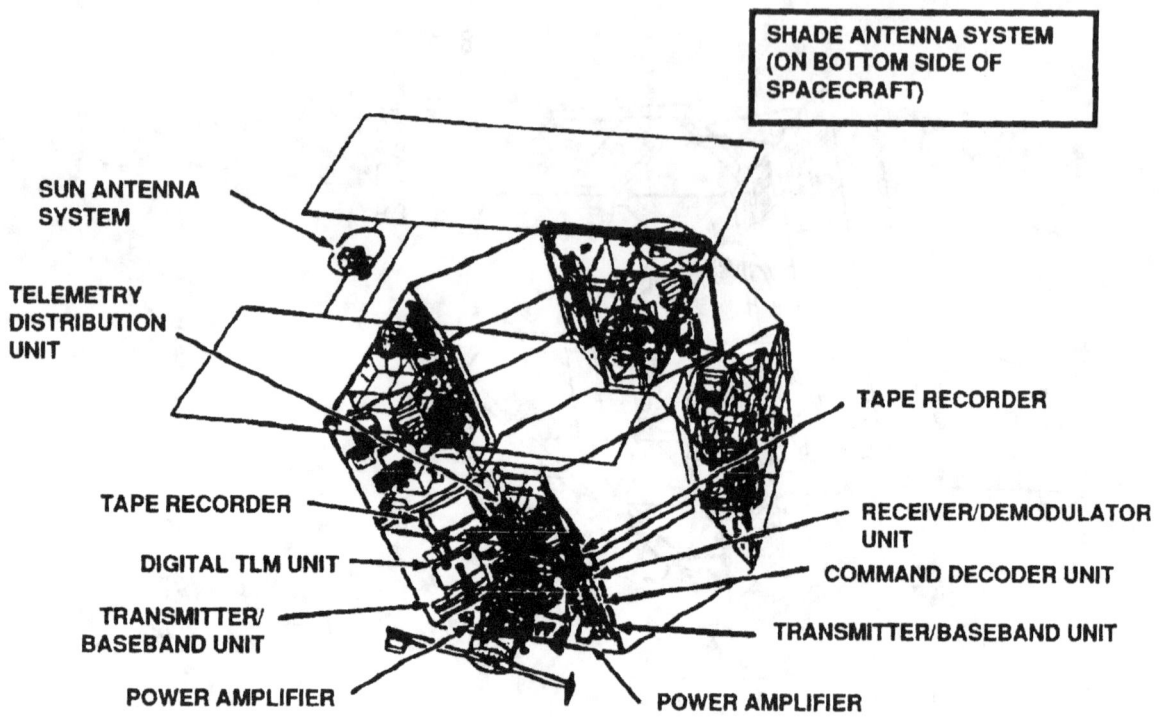

Perspective cutaway view of CRRES with attitude sensing and control equipment

Section 3 Operational Spacecraft

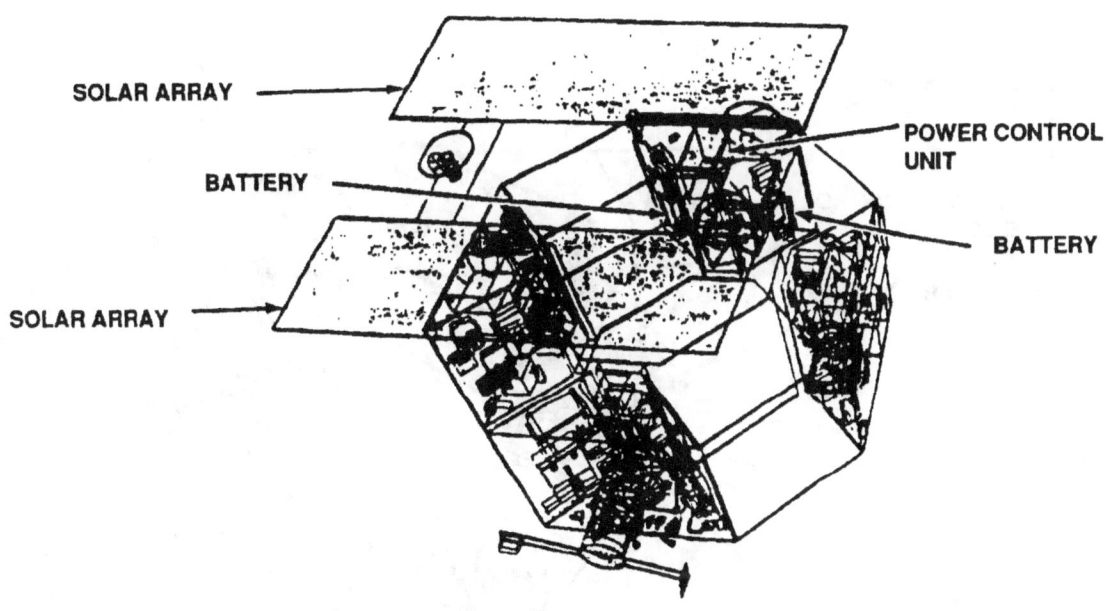

Perspective cutaway view of CRRES with attitude sensing and control equipment

Section 3 Operational Spacecraft

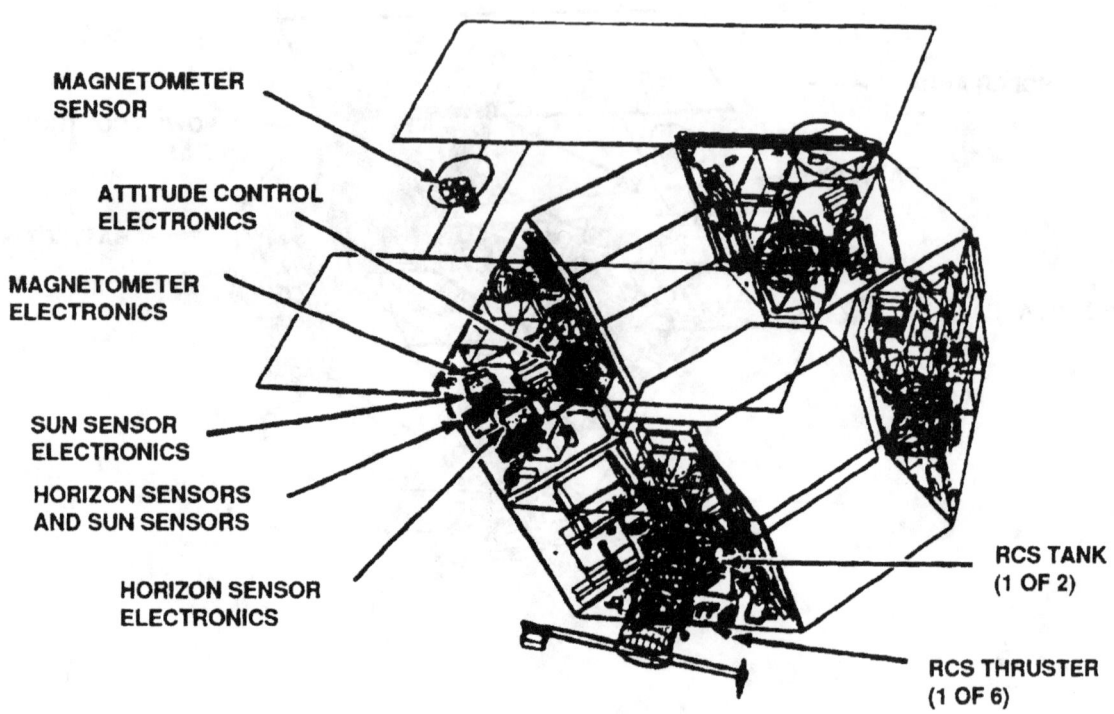

Perspective cutaway view of CRRES with attitude sensing and control equipment

Section 3 Operational Spacecraft

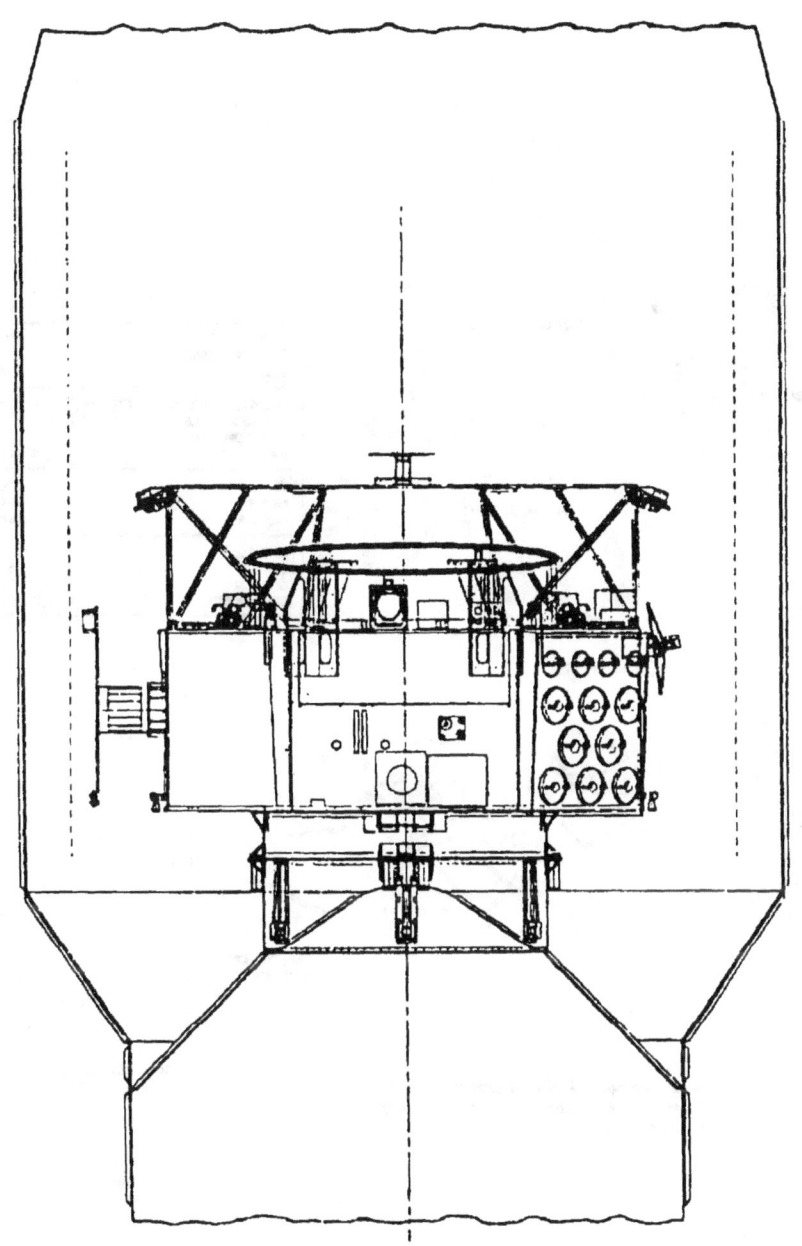

CRRES in launch configuration

Section 3 Operational Spacecraft

ATLAS I FLIGHT PROFILE

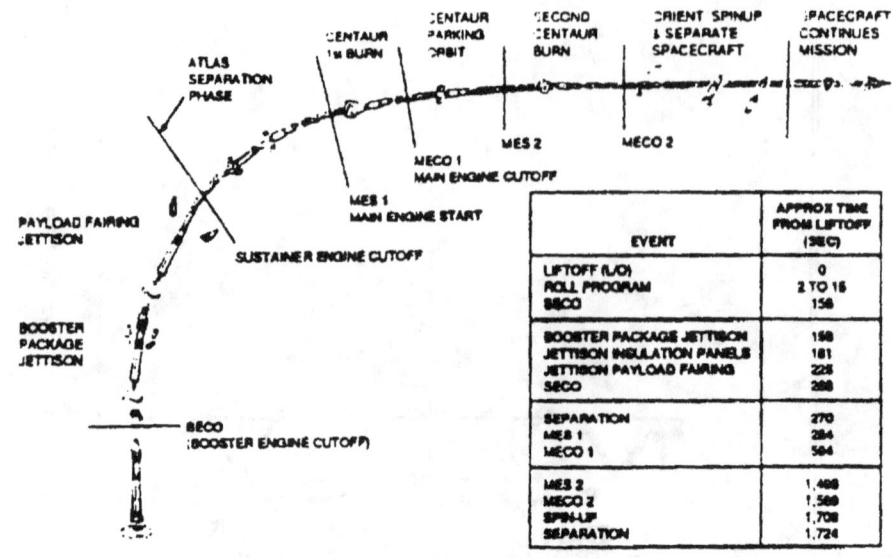

ON-ORBIT DEPLOYMENTS

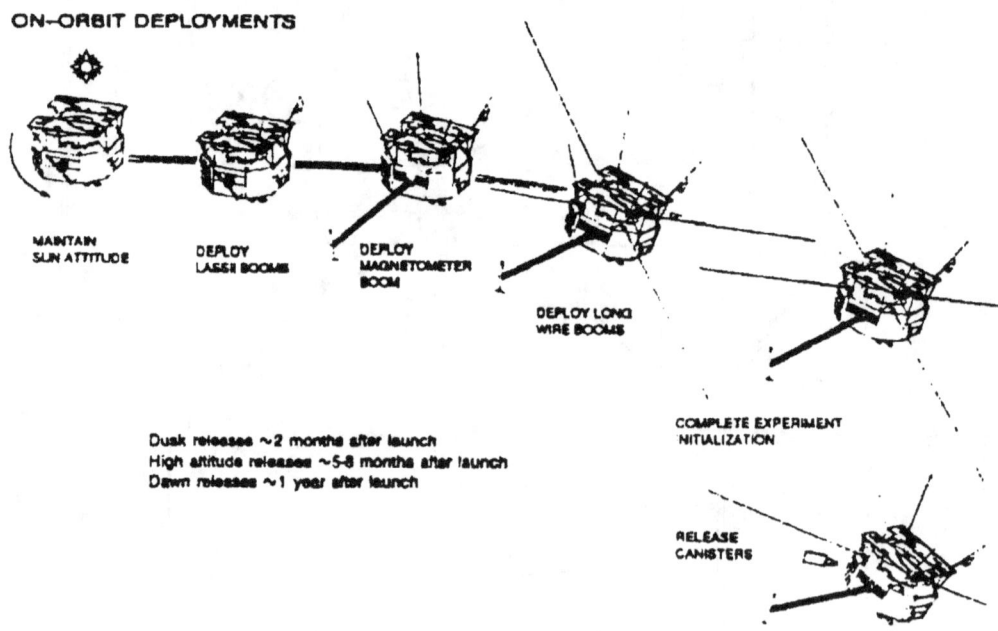

Dusk releases ~2 months after launch
High altitude releases ~5-8 months after launch
Dawn releases ~1 year after launch

CRRES launch profile

Section 3 Operational Spacecraft

CRRES satellite chemical release experiments

EXPERIMENT	Release Number	CANISTER TYPE and CHEMICAL	CHEM Wt.(kg)	RELEASE LOCATION	RELEASE ALTITUDE	RELEASE SEQUENCE	RELEASE PERIOD	RELEASE CONDITIONS	OBSERVATIONAL SUPPORT
Critical Velocity									
Critical Velocity Ionization	G-13	Large/Sr / Large/Ba	10 / 12.1	South Pacific (American Samoa)	450-600 km	First / 2.5 sec later	Aug. 1990	moon down, dusk 1900 LT	Two aircraft required Ground sites in Samoa and Fiji
	G-14	Large/Ca / Large/Ba	9.5 / 12.1	South Pacific (American Samoa)	450-600 km	First / 2.5 sec later	Aug. 1990	moon down, dusk 1900 LT	
High-Altitude Magnetospheric									
Diamagnetic Cavity, Plasma Coupling	G-1 / G-2 / G-3 / G-4	Small Ba,2% Sr / Small Ba,2% Sr / Small Ba,2% Sr / Small Ba,2% Sr	3.4 / 3.4 / 3.4 / 3.4	North America / North America / North America / North America	1.3 Re / 1.8 Re / 3.5 Re / 5.5 Re	n/a / n/a / n/a / n/a	N. Hemisphere Winter 1991	moon down	Northern Canada-NASA DC-8 and AFGL KC-135 Northern Pacific-Learjet South America-Argentine 707 Ground sites in both northern and southern hemispheres
Stimulated Electron Precipitation/Aurora Prod.	G-5	2 Large Li, 3% Eu	18.5	North America	>6.0 Re	both same time	N. Hemisphere Winter, 91	moon down 0000-0200 LT	Same as G-1, -2, -3, -4 with addition of Millstone Hill radar (foot of field)
Stimulated Ion-Cyclotron Waves and Ion Precipitation	G-6	2 Large Li, 3% Eu	18.5	North America	>6.0 Re	both same time	N. Hemisphere Winter, 91	moon down 2200-2400 LT	Same as G-1, -2, -3, -4 with addition of Millstone Hill radar (foot of field)
Ion Tracing and Acceleration	G-7	2 Large Li, 3% Eu	18.5	North America	>6.0 Re	both same time	N. Hemisphere Winter, 91	moon down 2000-2400 LT	Same as G-1, -2, -3, -4 with addition of Millstone Hill radar (foot of field)
Stimulating a Magnetospheric Substorm	G-10	2 Large Ba, 4% Li	24.8	North America	>6.0 Re	both same time	N. Hemisphere Winter, 91	moon down 0000-0200 LT	Ground sites in both northern and southern hemispheres
Caribbean Perigee									
Gravitational Instability, Field Equipotentiality	G-8	2 Large Ba, 2% Sr	24.4	field line pass thru Jicamarca, Peru (Grand Cayman Is.)	450-800 km	both same time	N. Hemisphere Summer, 91	moon down, dawn 0500 LT	NASA DC-8 and Argentine 707 ground sites in Ecuador, Dom. Rep. Jicamarca radar
Field Line Tracing and Equipotentiality	G-9 / G-11A / G-11B / G-12A / G-12B	2 Large Ba, 2% Sr / Small Ba,2% Sr / Small Ba,2% Sr / Small Ba,2% Sr / Small Ba,2% Sr	24.8 / 3.4 / 3.4 / 3.4 / 3.4	Caribbean latitudes / Caribbean latitudes / Caribbean latitudes / Caribbean latitudes / Caribbean latitudes	450-800 km / 450-800 km / 450-800 km / 450-800 km / 450-800 km	both same time / n/a / n/a / n/a / n/a	N. Hemisphere Summer, 91	two successive moon down periods, dawn 0500 LT	NASA DC-8 and AFGL KC-135 in Caribbean, Argentine 707 and Learjet in South America Arecibo radar and Jicamarca radar

* CHEMICAL WEIGHT INCLUDES THERMITE CHEMICALS

RELEASE #	PRINCIPAL INVESTIGATOR
G-1, G-2, G-3	HOFFMAN
G-4, G-6	MENDE
G-5	BERNHARDT + HAERENDEL
G-7	PETERSON
G-8	HAERENDEL
G-9	PONGRATZ + WESCOTT
G-10	SIMONS
G-11, G-12	WESCOTT
G-13, G-14	WESCOTT

3.7 Ulysses

Target: Solar polar regions

Orbit: Solar polar orbit (>70°), 1.28 AU x 5 AU, 5 year period

Mission Duration: 4.8 years

Mission Class: Moderate

Mass: 370 kg (total)
55 kg (payload)
33.5 kg (hydrazine)

Launch Vehicle: STS with IUS and PAM-S upper stages (5 October 1990)

Theme: Solar physics

Science Objectives:

The Ulysses mission will study the interplanetary medium and solar wind in the inner heliosphere as a function of heliographic latitude. It will for the first time permit *in situ* measurements to be made away from the plane of the ecliptic and over the polar regions of the Sun. The primary scientific objectives of the Ulysses (ULS) mission are to investigate, as a function of solar latitude, the properties of the solar corona, the solar wind, the structure of the Sun-wind interaction, the heliospheric magnetic field, solar and nonsolar cosmic rays, solar radio bursts and plasma waves, and the interstellar/interplanetary physics during the ecliptic phase and measurements of the Jovian magnetospheric during the Jupiter fly-by phase.

Spacecraft:

Type: Free flyer, ESA design, spin stabilized (5 rpm) RTG, 285 W
Telemetry: 1024 bps real time, 512 bps storage

Special Features: Radial dipole antenna—72.5 m
Axial monopole antenna—7.5 m
Radial magnetometer boom—5.6 m

Special Requirements: None

Instruments:

Experiment	Principal Investigator	Exper. code	Measurement	Instrumentation
Magnetic Field	A. Balogh (Imperial College, London)	HED	Spatial and temporal variations of the heliosphere and Jovian magnetic field in the range ±0.01 nT to ±44000 nT.	Triaxial vector helium and fluxgate magnetometers.
Solar Wind Plasma	S.J. Bame (Los Alamos National Lab.)	BAM	Solar-wind ions between 257 eV and 35 keV. Solar-wind electrons between 1 eV and 903 eV.	Two electrostatic analyzers with channel electron multipliers (CEMs).
Solar-Wind Ion Composition	G. Gloeckler (Univ. Maryland); J. Geiss (Univ. Bern)	GLG	Elemental and ionic-charge composition, temperature and mean velocity of solar-wind ions for speeds from 145 km/s (H^+) to 1352 km/s (Fe^{14}).	Electrostatic analyzer with time-of-flight and energy measurement.
Low-Energy Ions and Electrons	L. Lanzerotti (Bell Laboratories)	LAN	Energetic ions from 50 keV to 5 MeV. Electrons from 30 keV to 300 keV.	Two sensor heads with five solid-state detector telescopes.
Energetic-Particle Composition and Interstellar Gas	E. Keppler (Max-Planck-Institut, Lindau)	KEP	Composition of energetic ions from 80 keV to 15 MeV per nucleon. Interstellar neutral helium.	Four solid-state detector telescopes. LiF-coated conversion plates with CEMs.
Cosmic Rays Solar Particles	J.A. Simpson (Univ. Chicago)	SIM	Cosmic rays and energetic solar particles in the range 0.3–600 MeV per nucleon. Electrons in the range 4–2000 MeV.	Five solid-state detector telescopes. One double Cerenkov and semi-conductor telescope for electrons.
Unified Radio and Plasma Waves	R.G. Stone (NASA GSFC)	STO	Plasma waves: remote sensing of travelling solar radio bursts and electron density. Electric field: Plasma waves: 0–60 kHz Radio receiver: 1–940 kHz Magnetic field: 10–500 Hz	72m radial dipole antenna 7.5m axial monopole antenna. Two-axis search coil.
Solar X-Rays and Cosmic Gamma-Ray Bursts	K. Hurley (Berkeley); M. Sommer (MPI)	HUS	Solar-flare X-rays and cosmic gamma-ray bursts in the energy range 5–150 keV.	Two Si solid-state detectors. Two CsI scintillation crystals.
Cosmic Dust	E. Grün (MPI, Heidelberg)	GRU	Direct measurement of particulate matter in mass range 10^{-7}–10^{-16} g.	Multi-coincidence impact detector with channeltron.
Coronal Sounding	H. Volland (Univ. Bonn)	SCE	Density, velocity, and turbulence spectra of solar coronal plasma.	Spacecraft-to-Earth dual ranging doppler data.
Gravitational Waves	B. Bertotti (Univ. Pavia)	GWE	Search for gravitational waves.	Spacecraft two-way doppler data.

Mission Strategy:

Ulysses is a joint mission between NASA and ESA to explore the heliosphere over the full range of solar latitudes, and provide an accurate assessment of our total solar environment. The spacecraft was launched on 5 October 1990 by the shuttle *Discovery*, and a PAM/IUS subsequently placed the spacecraft on an ecliptic transfer orbit to Jupiter. At the time of launch Ulysses was the fastest man-made object in the Universe, with an escape velocity of 11.4 km/s. On its outward journey Ulysses will perform interplanetary physics investigations and, during the two periods when Earth is directly between the spacecraft and the Sun, searches for gravitational waves will be conducted. Upon arrival at Jupiter, in February 1992, Ulysses will measure the Jovian magnetosphere. The massive gravitational field of the planet will then be used to deflect the spacecraft trajectory downward and away from the ecliptic plane, placing Ulysses in a high-inclination orbit towards the Sun's south pole. The most intense scientific activity will commence with the first south polar pass, in May 1994. As the spacecraft passes through 70° south solar latitude it will collect information about the solar corona, the solar wind, the heliospheric magnetic field, cosmic rays, and other particles and fields. The spacecraft will spend about four months in that region, at a distance of 330 million kilometers (200 million miles) from the Sun. In response to the Sun's gravitational pull, Ulysses will then arc towards the solar equator and cross it in February 1995 at about 1 AU. Though scientific investigations will continue throughout this period, activity will peak again when Ulysses begins a four-month pass of the Sun's north polar region in May 1995. By September 1995, about five years after launch, Ulysses will have explored the heliosphere at nearly all latitudes, measured phenomena over both of the Sun's poles, and investigated interplanetary particles and fields. As the spacecraft leaves the region of the Sun's north pole, the mission ends.

Enabling Technology Development: None

Points of Contact:

Program Manager:	R.F. Murray
Program Scientist:	Dave Bohlin
NASA Center:	JPL
Project Manager:	W. Weeks
Project Scientist:	Ed Smith
ESA Project Manager:	D. Eaton
ESA Project Scientist:	K. Wenzel

Section 3 Operational Spacecraft

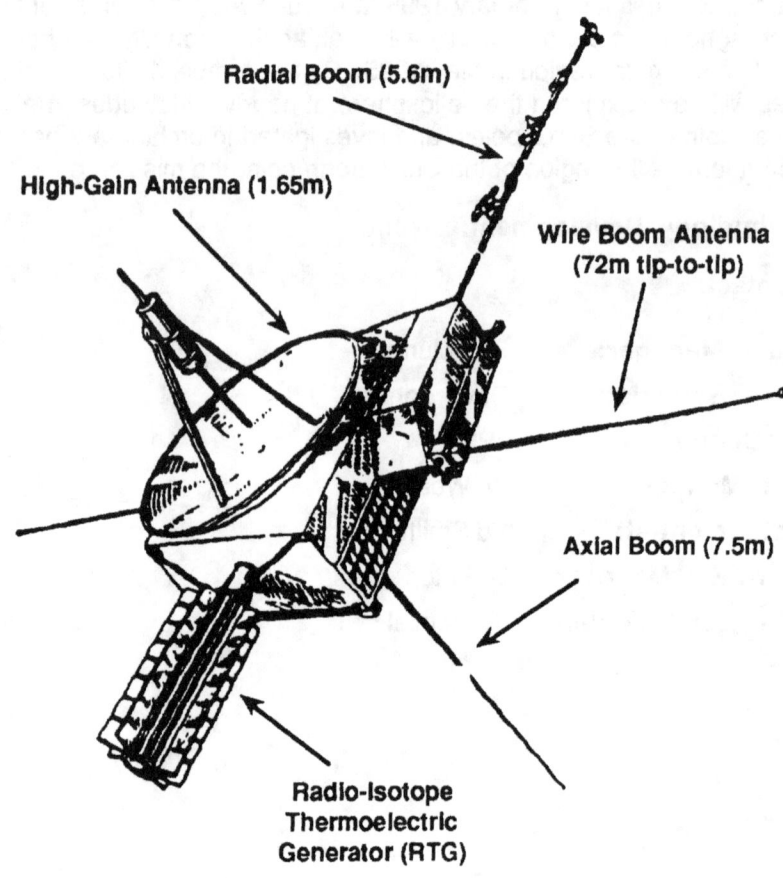

Ulysses spacecraft in-flight configuration

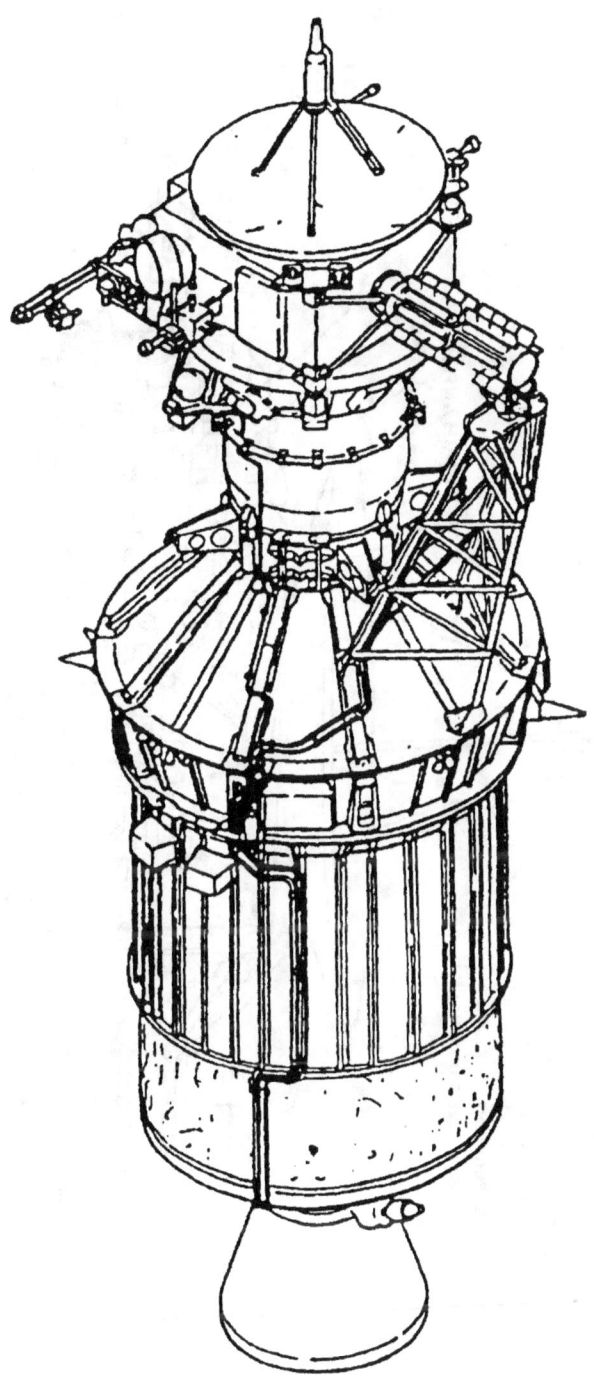

Ulysses

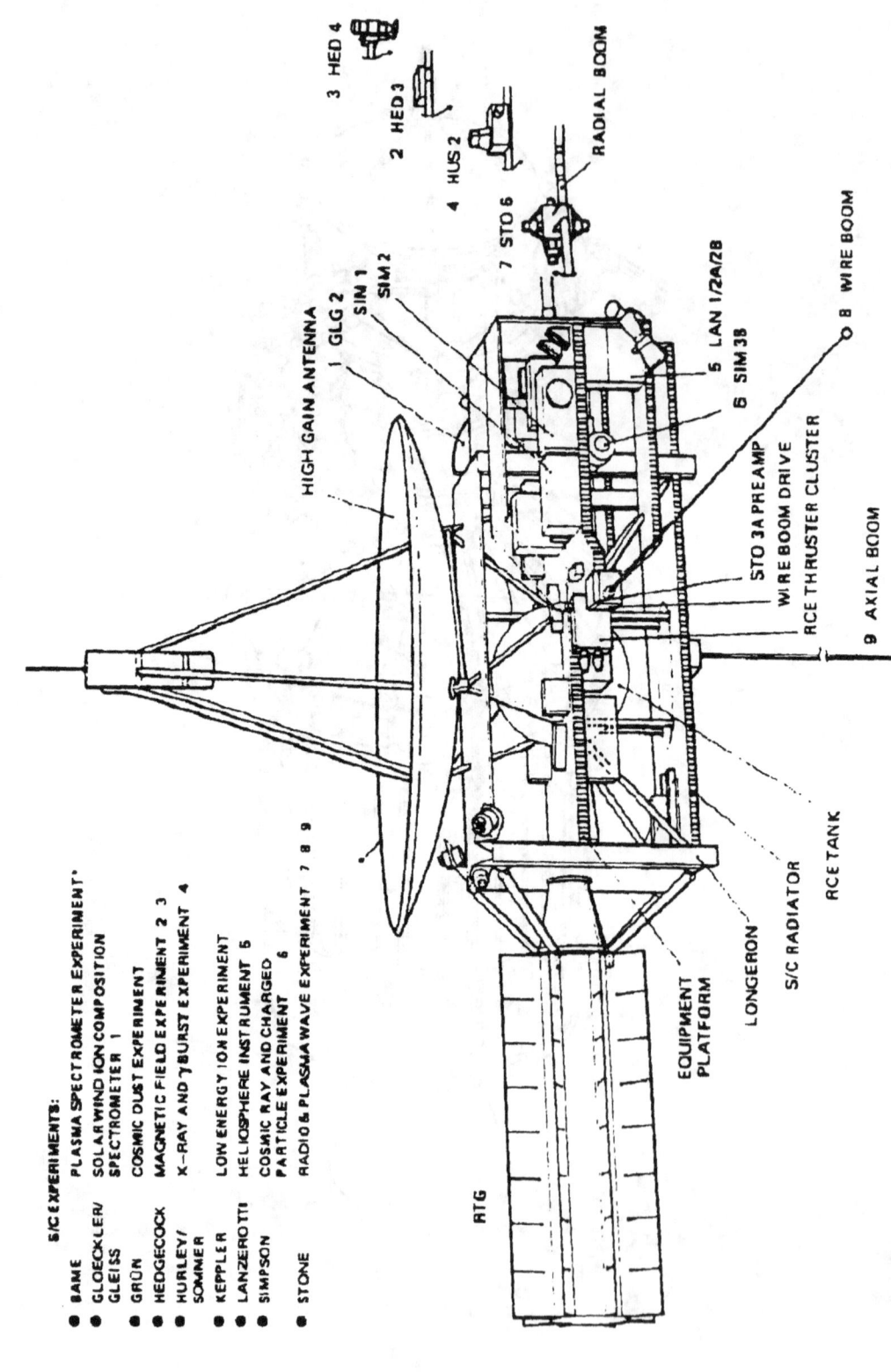

Ulysses spacecraft configuration

Section 3 Operational Spacecraft

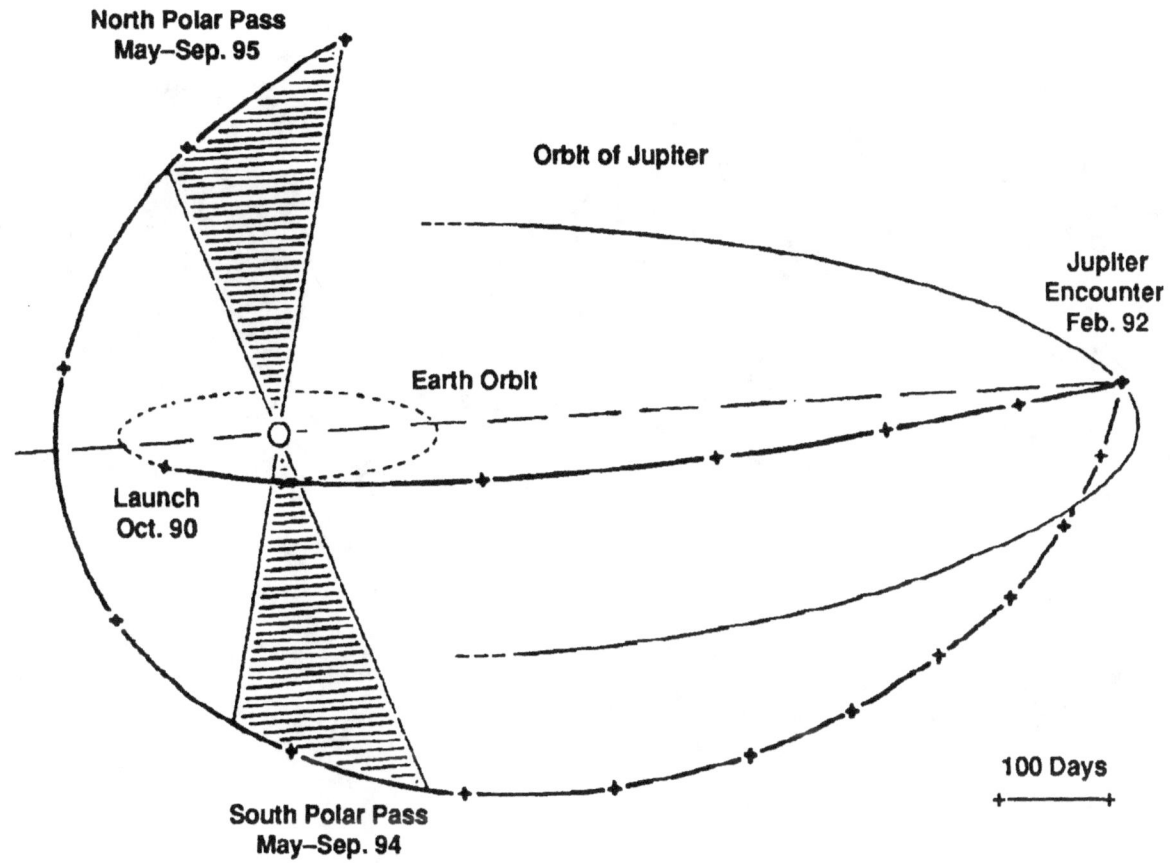

**Typical Ulysses spacecraft trajectory viewed
from 15° above the eliptic plane**

3.8 Sounding Rocket Program

Target: Various

Orbit: Suborbital
    Max. altitude: 1800 km (Black Brant XII)

Mission Duration: 22 minutes max.

Mission Class: Suborbital program

Mass: 950 kg max (Aries)

Launch Vehicle: NASA sounding rockets

Theme: Space science

Science Objectives:

Various space science objectives. Most relate to ionospheric-thermospheric-mesospheric or solar physics.

Spacecraft:

    Type: TBD

    Special Features: Reusable bus, parachute recovery system, mother/daughter payloads, etc.

    Special Requirements: Rocket type and launch location chosen to match science requirements

Instruments: Various

Mission Strategy:

The NASA Sounding Rocket Program is a suborbital space flight program primarily in support of space and Earth sciences research activities sponsored by NASA. This program also provides applicable support to other government agencies as well as to international sounding rocket groups and scientists.

Since the program's beginning in 1959, there have been almost 2,500 flight missions with a science mission success of over 85 percent. Launch vehicle success rate is over 95 percent. The program is a low-cost, quick-response effort that currently provides approximately 40 flight opportunities per year to space scientists involved in the disciplines of upper atmosphere, plasma physics, solar physics, planetary atmospheres, galactic astronomy, high-energy astrophysics, and micro-gravity research. These rockets are launched from a variety of launch sites throughout the free world.

Although this program is conducted without the formal and expensive reliability and quality assurance employed in the larger and more costly orbital and deep space programs, overall mission success reliability is over 85%. This informal approach, and the extensive use of military surplus motors (11 of 13 launch vehicles), is instrumental in enabling the support of approximately 40 missions per year with the available NASA resources.

A family of standard sounding rocket launch vehicles is available in the NASA Sounding Rocket Program for use in conducting suborbital space science, upper atmosphere and other special applications research. Some of the vehicles are commercially available; others have been developed by NASA for exclusive use in NASA programs. These vehicles are capable of accommodating a wide variety of payload configurations and providing an extensive performance envelope.

There are currently thirteen operational support launch vehicles in the NASA Sounding Rocket Program and an additional two under development. All NASA sounding rocket launch vehicles use solid propellant propulsion systems. Extensive use is made of 20 to 30 year old military surplus motors in 11 of the systems. All vehicles are unguided except the Aries and those which use the S-19 Boost Guidance System. During flight all launch vehicles, except the Aries, are imparted with a spinning motion to reduce potential dispersion of the flight trajectory due to vehicle misalignments.

Enabling Technology Development: None

Points of Contact:

Program Manager:	Glenn Mucklow
Program Scientist:	Dave Evans
NASA Center:	GSFC (WFF)
Project Manager:	Larry Early (WFF)
Project Scientist:	Werner Neupert (GSFC)

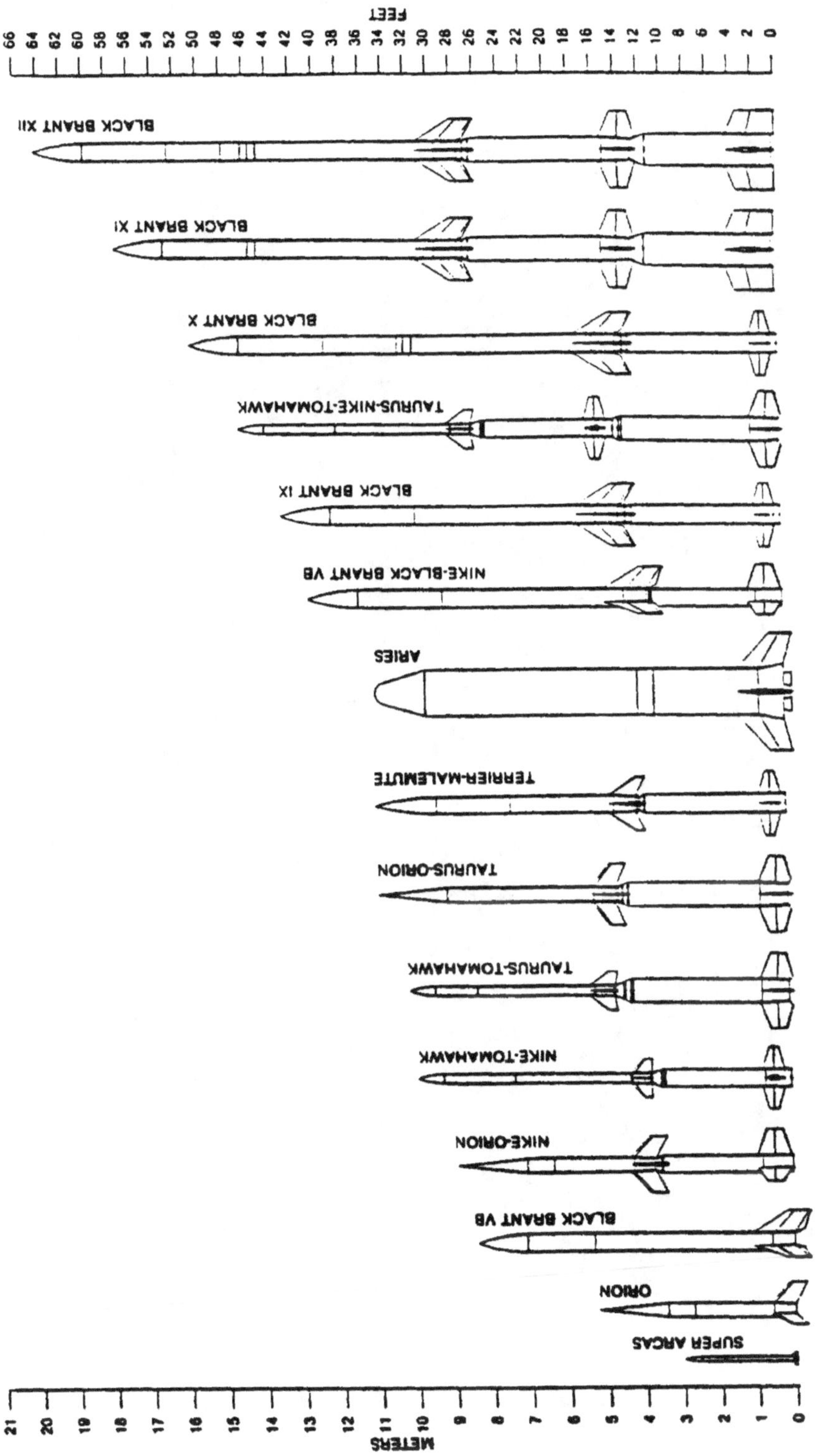

Section 3 Operational Spacecraft

NASA sounding rockets

3.9 Scientific Balloon Program

Target: Various

Orbit: Suborbital
   Max altitude: 43 km (680 kg)

Mission Duration: Days to weeks

Mission Class: Suborbital program

Mass: 3400 kg (max at 30 km alt.)

Launch Vehicle: Scientific balloons

Theme: Space science

Science Objectives:

Various solar, cosmic and heliospheric, and other space science

Spacecraft:

   Type: Scientific balloon

   Special Features: Balloon type and launch location chosen to match science requirements

   Special Requirements: Varies with science

Instruments:

Various

Mission Strategy:

The NASA Balloon Program conducts approximately 50 high-altitude balloon flights per year, predominantly in support of the astrophysics, space physics, and upper atmosphere research programs. The flights, which are used for both fundamental science investigations and for tests of hardware for future space missions, have generally been carried out from the National Scientific Balloon Facility (NSBF) at Palestine, Texas. Remote launch sites have traditionally been used to accomplish specific scientific objectives, e.g., those requiring the low geomagnetic cutoff at high geographic latitudes or celestial targets not observable from Palestine. Flights requiring more than several hours of observations are typically carried out during the so-called "turnaround" seasons, those few weeks during both the spring and fall when high-altitude winds are slow and variable because the prevailing east-west high-altitude wind direction is reversing.

Section 3 Operational Spacecraft

The current balloon program relies on five standard balloons having the characteristics and capabilities given in Table 1. There are three basic balloon sizes, two of which have heavy (H) versions characterized by greater effective wall thickness, i.e., more caps. Only one of the balloons is designed for a nominal float altitude of 130,000 ft, although a range of altitudes can be accommodated within the minimum and maximum weight limits. The science payload weights that can be accommodated depend primarily on the flight time desired and the time of day of the launch. Payloads up to a few hundred pounds can be flown on much smaller balloons (e.g., 0.25–3 MCF; 1 MCF=10^6 ft^3) that are not included with the standard heavy-payload balloons listed in Table 1.

Balloon volume, MCF	Balloon weight, lb	Suspended, weight, lb	NSBF systems, lb	Nominal altitude, ft in 1000's	Flight[a,b] ballast science, lb		Flight[b,c] ballast science, lb	
					10h	12h	20h	36h
12	1710	2100	405	120	220	1475	695	1000
12H	3220	5500	650	100	525	4325	1575	3275
23	3750	4300	600	120	475	3225	1450	2250
28	3355	3000	500	130	375	2125	1150	1350
28H	4555	5500	650	120	600	4250	1825	3025

[a] 10–12-h flight assumes a morning launch. [b] Weights listed are nominal; the maximum and minimum payload weights would vary with the ballast allocation. [c] 20–36-h flight assumes a morning or evening launch.

Enabling Technology Development: None

Points of Contact:

Program Manager:	Glenn Mucklow
Program Scientist:	Vernon Jones
NASA Center:	GSFC (WFF)
Project Manager:	Harvey Needleman (WFF)
Project Scientist:	Bob Hartman (GSFC)

Section 3 Operational Spacecraft

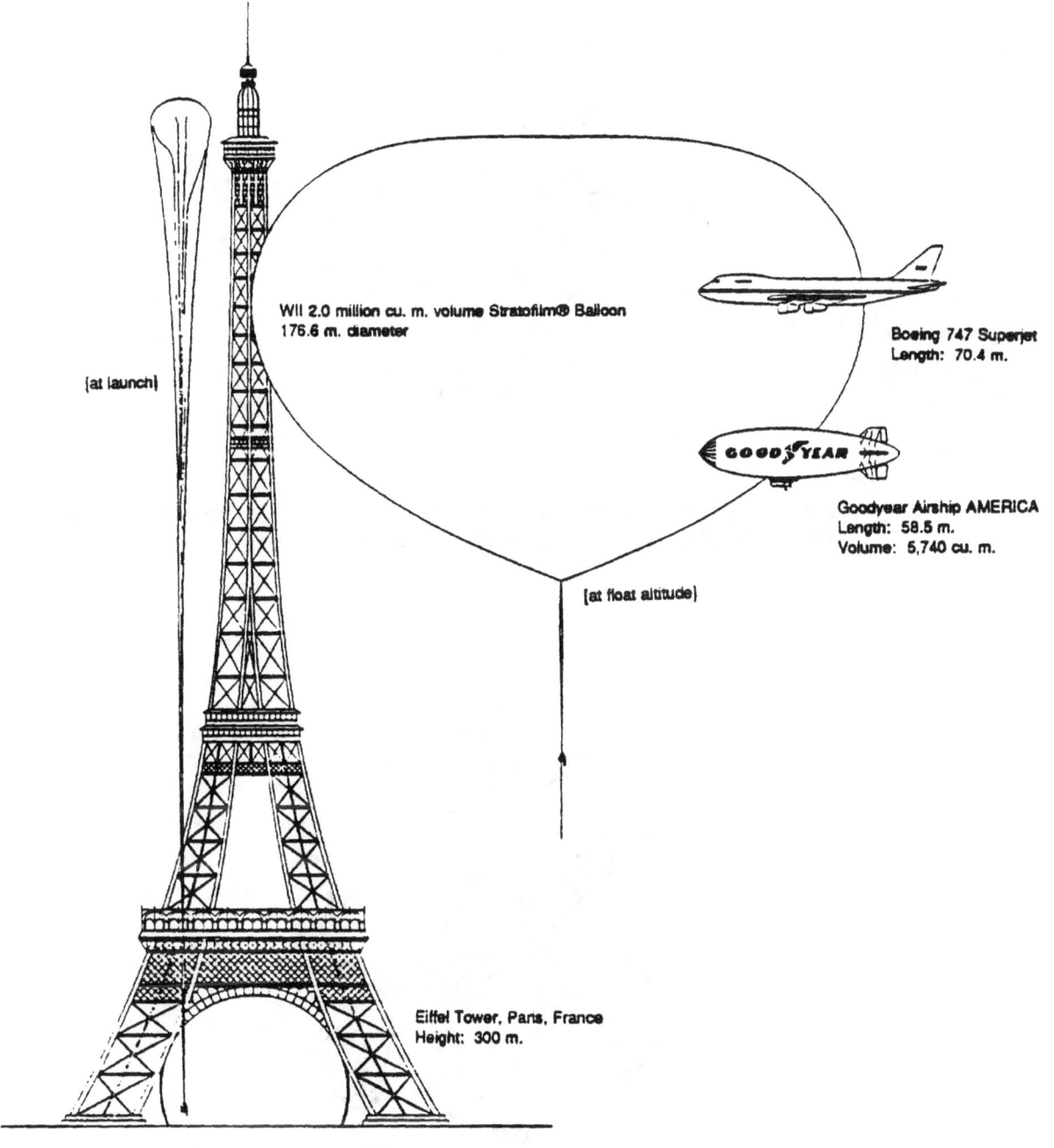

NASA balloon size

Section 3 Operational Spacecraft

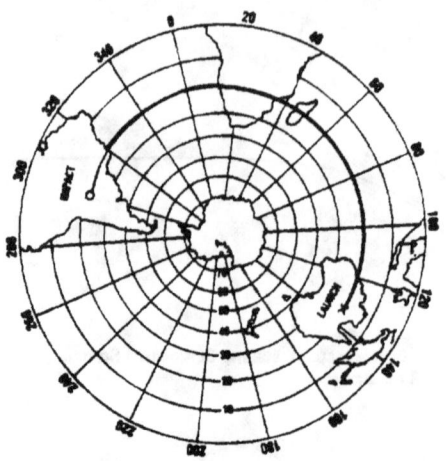

View from the South Pole of the trajectory of balloons launched from Australia and recovered in South America

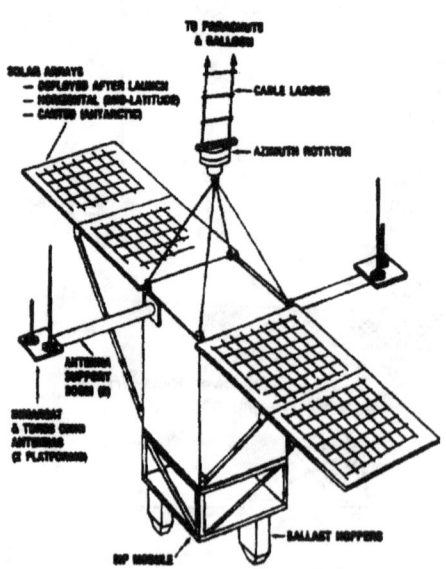

Schematic diagram of long-duration flight system

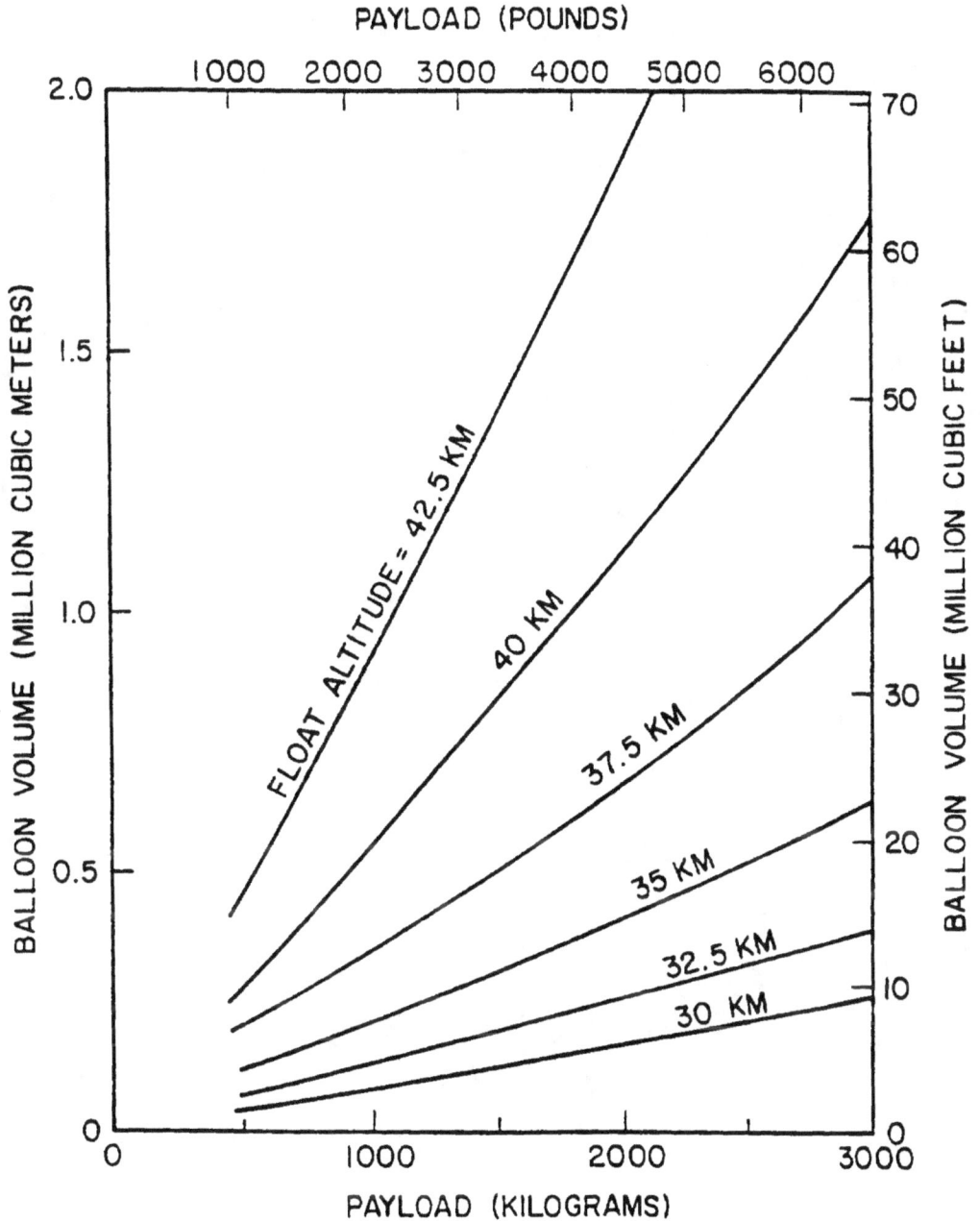

Balloon payloads and volumes

Section 4
Approved Missions

4.1 SAMPEX

Target: Solar energetic particles and cosmic rays

Orbit: 580 km x 580 km, 82° inclination

Mission Duration: 3 years nominal

Mission Class: Small Explorer

Mass: 200 kg

Launch Vehicle: Scout (June 1992)

Theme: Cosmic and heliospheric physics

Science Objectives:

The zenith-pointing satellite, in near-polar orbit, will carry out energetic-particle studies of outstanding questions in the fields of space-plasma physics, solar physics, heliospheric physics, cosmic-ray physics, and middle-atmospheric physics. The SAMPEX satellite will measure the composition of the following:

- Solar energetic particles
- Anomalous cosmic rays
- Galactic cosmic rays.

The designed mission lifetime of at least three years will allow the solar studies to be carried out from immediately after the peak of the solar maximum into the declining phase of the activity cycle. In particular it will be able to accomplish the following objectives:

- Measure the dependence of fluxes on the declining solar activity cycle.
- Determine flux levels and local time dependence of precipitating electrons during declining solar cycle.

Spacecraft:

> **Type:** Small Explorer, spin stabilized
> **Special Features:** Nadir pointer
> **Special Requirements:** TBD

Instruments:

Investigation	Expt code	Mass (kg)	Data rate (bps)	Principal Investigator
Low Energy Ion Composition Analyzer	LEICA	7.7	2000	Univ. of Maryland
Heavy Ion Large Telescope	HILT	24.0	2000	MPI
Mass Spectrometer Telescope	MAST	7.1	2000	CalTech
Proton/Electron Telescope	PET	(included with MAST)		CalTech

The payload is comprised of four particle detectors, each of which addresses a subset of the required measurements, and there is enough overlap in energy and response to allow inter-calibration and partial redundancy. Two of the instruments, the Low Energy Ion Composition Analyzer (LEICA) and the Heavy Ion Large Telescope (HILT), were originally designed and constructed for space shuttle flight together in the Getaway-Special program. The other two instruments, the Mass Spectrometer Telescope (MAST) and the Proton/Electron Telescope (PET), comprise a single instrument package that was designed to fly on the proposed U.S. spacecraft of the International Solar Polar Mission. These versatile instruments can provide isotope measurements over a broad energy range with outstanding resolution.

	LEICA	HILT	MAST	PET
Energy range				
Electrons (MeV)				01–120
H (MeV)	0.75–8		01–15	18–350
He (MeV/Nucleon)	0.41–8	3.9–9.0	07–91	18–500
C (MeV/Nucleon)	0.35–12	7.2–160	12–210	
Si (MeV/Nucleon)	0.26–18	9.6–117	19–345	54–195
Fe (MeV/Nucleon)	0.16–25	11–90	24–470	
Field of view (°)	24x20	90x90	101 cone	58 cone
Weight (kg)	7.7	24	7.1	(incl. w/MAST)
Power (W)	5.4	7	4.3	(incl. w/MAST)
Telemetry (kbps)	2	2	2	(incl. w/MAST)

Mission Strategy:

SAMPEX will be launched from Vandenberg AFB in June 1992, close to solar maximum. The spacecraft will sample particles and cosmic rays while pointing in a generally zenith direction, especially over the poles. The mission should be able to monitor at least one major solar flare event and several smaller to moderate solar flare events. The SAMPEX spacecraft will provide on-board storage of data, so it will not be be necessary to transmit data to a NASA ground station more frequently than once a day. The University of Maryland will process the raw data to assure correct timing and to remove overlaps, and then merge the science data with ephemeris/attitude data supplied by NASA.

Enabling Technology Development: None

Points Of Contact:

Program Manager:	John Lintott/SZ
Program Scientist:	Vernon Jones
NASA Center:	GSFC
Project Manager:	TBD
Project Scientist:	TBD

4.2 Solar-A

Target: Sun

Orbit: Circular, 550 km, 31° inclination

Mission Duration: 3 years

Mission Class: Small

Mass: 410 kg

Launch Vehicle: M-3SII-6 (Aug 1991)

Theme: Solar physics

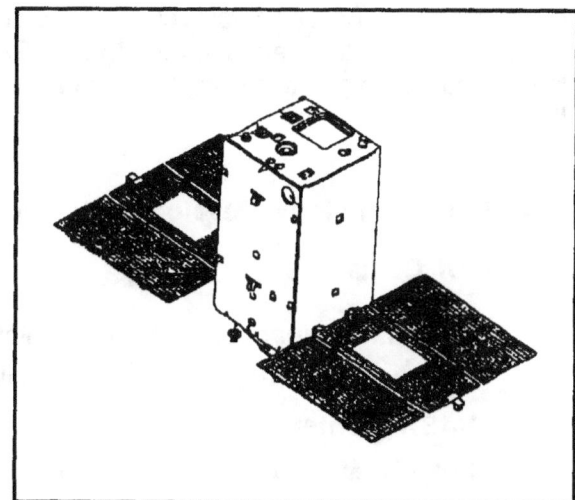

Science Objectives:

The primary goal of the Solar-A mission is to investigate high-energy phenomena on the Sun through X-ray and gamma-ray observations taken by a carefully coordinated set of instruments. Solar-A will, for the first time, observe the soft X-ray image and the hard X-ray image of solar flares simultaneously. The mission will also observe the energy spectrum over a wide energy range.

Spacecraft:

    Type: Special ISAS Design. 3-axis stabilized

    Special Features: TBD

    Special Requirements: Sun pointing, Sun angle less than 1.5°
                                   Antenna pointing accuracy—6 arc-minutes

Instruments:

Hard X-ray telescope	48 kg	Multi-pitch bi-grid modulation collimators with Na I scintillators
Soft X-ray telescope	27 kg	Grazing incidence mirror with CCD detector
Bragg crystal spectrometer	13 kg	Bent crystals with position sensitive proportional counters
Wide-band spectrometer	16 kg	Proportional counter, Na I scintillator, and two BGO scintillators

Mission Strategy:

Solar-A is a joint mission with the Japanese ISAS, and will be launched from Kagoshima Space Center. The Soft X-ray Telescope (SXT) investigation and the Bragg crystal spectrometer (BCS) investigation are shared with a NASA-sponsored U.S. team of co-investigators. A NASA-sponsored guest investigator program with access to all mission data will be started in fiscal 1994.

Enabling Technology Development: None

Points Of Contact:

Program Manager:	John Lintott
Program Scientist:	Dave Bohlin
NASA Center:	MSFC
Project Manager:	Rein Ise
Project Scientist:	John Davis

Section 4 Approved Missions

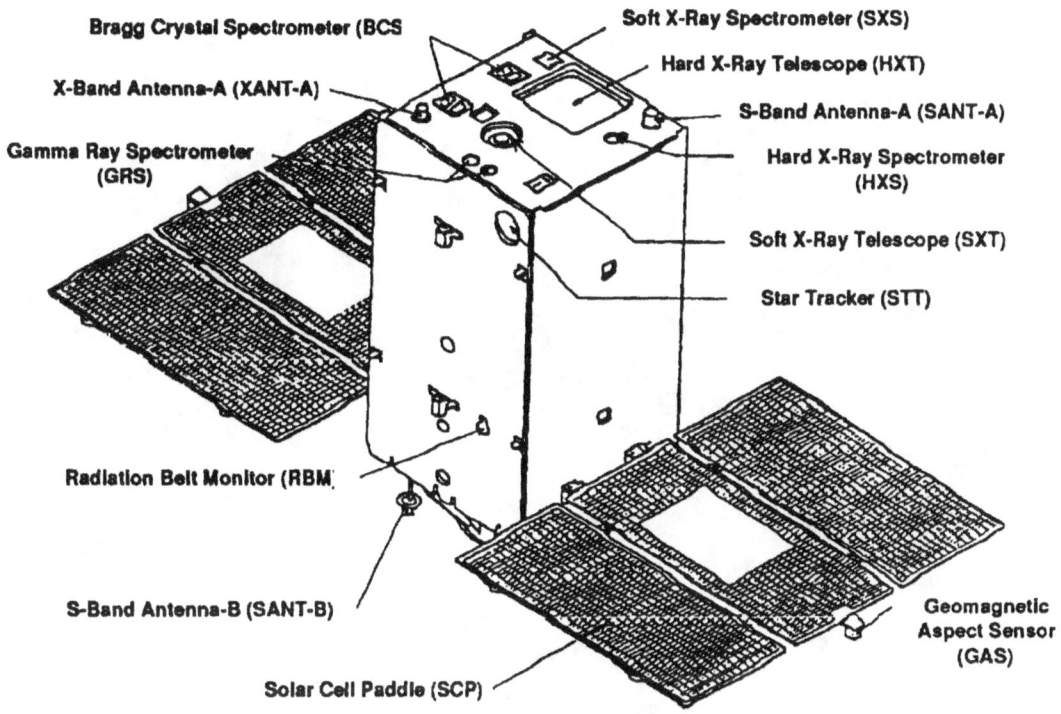

Solar-A

4.3 Tethered Satellite System (TSS)

Target: Electrodynamic tether/ambient plasma interactions

Orbit: 297 km altitude, 28.5° inclination

Mission Duration: 36 hours

Mission Class: Small

Mass: TBD

Launch Vehicle: STS (Jan 1991)

Theme: Ionospheric physics

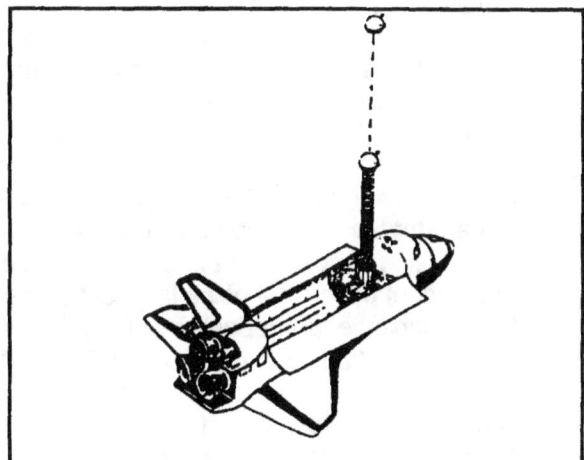

Science Objectives:

The objectives of this first tethered satellite mission are as follows:

- Investigate interactions between the tether/satellite/shuttle and the ambient plasma.
- Study the physics of electrodynamic tethers in space.
- Investigate the dynamics of tethered systems.
- Demonstrate the capability to deploy, maintain, and retrieve a tethered satellite.

Spacecraft:

> **Type:** STS, tethered payload
> **Special Features:** TBD
> **Special Requirements:** TBD

Instruments:

Ten scientific investigations were selected for the first flight on the TSS. Experimental work will be conducted using instrumentation on the satellite, on the Orbiter, and on the ground. Theoretical studies will support the experimental efforts. The investigations are:

- *Shuttle Electrodynamic Tether System (SETS).* The SETS experiment will provide a fast-pulsed electron generator (FPEG) and three diagnostic instruments—a spherical retarding potential analyzer (SRPA), a spherical Langmuir probe (SLP), and a charge-and-current probe (CCP). It will also provide tether current and voltage measurements. Each of these instruments will be mounted in the cargo bay. The FPEG will be referenced directly to the Orbiter's ground bus through a high-voltage accelerator supply, and can emit 50 and 100 mA electron beams at an energy of 1 keV. The emitted beams can be pulse modulated with on-off times ranging from 600 ns to 105 s, with a selectable number of pulses in a given pulse train. The SRPA will measure the ion current density, energy, and temperature; the SLP will measure the local space potential and the number, density, and temperature for electrons; and the CCP will provide a measurement of the return current to the Orbiter. (P.M. Banks, Stanford University)

- *A Theoretical and Experimental Investigation on TSS Dynamics.* This investigation will attempt to verify the basic models for tethered satellite deployment and retrieval mechanics, observe the effects of dynamic and electrodynamic drag, and determine the level of dynamical noise inherent to the system. This last question is of particular significance to the design of future TSS-borne experiments to study the small-scale gravitational structure of the Earth. Accelerometer data from the TSS satellite will be analyzed to verify TSS' capability of resolving dynamical effects with time scales down to less than 1 Hz, especially damping techniques that can be used on future missions to obtain acceptable dynamic noise levels for gravitational investigations. In addition, these investigations will address the response of the TSS to know impulses, and the mechanical characteristics of the tether itself. (S. Bergamaschi, University of Padova, Italy)

- *Investigation of Dynamic Noise in TSS.* This effort complements the previous one. (G. Gullhorn, Smithsonian Astrophysical Observatory)

- *Italian Core Equipment for Tether Current/Voltage Control.* The Orbiter-mounted core equipment will consist of a core electron generator (CEG) and a set of high-voltage tether isolation switches. The CEG is capable of producing electron beams of up to).5 Amp at 5 kV. Its cathode is connected directly to the tether, and the emitted electrons are accelerated with the tether's dynamically generated potential, which is expected to reach about 5 kV at maximum deployment. The high-voltage tether isolation switches will be capable of operating while the tether current is flowing and will be used to connect the Orbiter end of the tether either directly to the CEG cathode or to the Orbiter ground bus through the SETS experiment. The satellite-mounted equipment will include high-voltage switches to allow direct connection to the satellite ground bus (and surface), a triaxial accelerometer to determine satellite dynamical characteristics, and an ammeter to measure the tether current. (C. Bonifazi, CNR, Italy)

- *Research on Electrodynamic Tether Effects (RETE).* The RETE experiment will provide a pair of 5 m extendable booms that will carry a triaxial AC electric field meter, a search coil magnetometer, and a Langmuir probe. The Langmuir probe will determine the temperature and density of electrons and the local space potential with respect to the satellite potential. The AC electric field will be measured from dc to 200 Hz by two Langmuir probe sensors and from 200 Hz to 2 MHz with the electric field meter. The search coil magnetometer will measure the magnetic field intensity over the same frequency range. (M. Dobrowolny, CNR, Italy)

Section 4 Approved Missions

- *Theory and Modeling in Support of Tether.* This investigation will develop and test theories and models for the interaction of the satellite/tether/orbiter system with the plasma environment. Both steady-state and time-varying conditions will be considered and related to fundamental processes in space plasma physics. (A. Drobot, Science Applications International Corporation)

- *Investigation of Electromagnetic Emissions from Electrodynamic Tether.* This program for ground-based measurement of electromagnetic emissions excited by the TSS will make use of several facilities, including extremely low frequency (ELF) radio receivers at the Arecibo facility, magnetometers used as ELF receivers at several locations of the worldwide geomagnetic observatories chain, and an Italian ocean surface and ocean bottom observational facility. The ground-based facilities will provide measurements of the electromagnetic emissions, with frequencies ranging from dc to about 3 kHz. (R. Estes, Smithsonian Astrophysical Observatory)

- *Detection of the Earth's Surface of ULF/VLF Emissions by TSS.* This effort extends the above investigation to ultra-low and very-low frequencies. (G. Tacconi, University of Genoa, Italy)

- *Magnetic Field Experiment for the TSS Missions (TMAG).* The TMAG experiment will include two triaxial fluxgate magnetic field sensors mounted on a satellite boom. This instrument will assess the relative influence of the field inherent to the satellite and will measure the magnetic signature created by the presence of the rapidly moving, highly biased satellite in the ionospheric magnetoplasma. The magnetic field vector will be measured 15–20 times and with a resolution of ± 1 nT. (F. Mariani, the 2nd University of Rome and CNR, Italy)

- *Research on Orbital Plasma Electrodynamics (ROPE).* The ROPE experiment will provide charged particle measurements at the surface of the satellite and on a satellite boom. The boom-mounted differential ion flux probe (DIFP) will measure the current density, flow direction, energy, and temperature of ions passing around the satellite or created within its plasma sheath and accelerated outward. The DIFP has an energy range of 0–1000 eV with 10 percent accuracy. The ion stream angle-of-attack is measured over an angular range of ±45 degrees with ±1 degree accuracy. The soft particle energy spectrometer (SPES) will measure the energy distribution of both electrons and ions, over the range 10 eV to 10 KeV, at seven locations. The outward and inward streaming ions and electrons will be measured at six locations on the satellite's surface. (N. H. stone, Marshall Space Flight Center)

Mission Strategy:

The TSS is a cooperative program of NASA and the National Research Council (NRC) of Italy. NASA is responsible for the TSS deployer and systems integration; Italy is building the satellite; and both are providing scientific investigations. On its first flight, the satellite will be deployed on a conducting tether above the orbiter to a distance of 20 km. The deployer consists of a Spacelab pallet, a reel for tether deployment, an extendible/retractable boom for initial deployment and final retrieval of the satellite, an electrical power and distribution subsystem, a communications and data management subsystem, and a tether control capability. A separate support structure will carry science instrumentation. The satellite is spherical with a diameter of 1.5 meters. The upper hemisphere will house the scientific payload, while the lower hemisphere contains the support equipment. The satellite is equipped with cold-gas (nitrogen) thrusters used for deployment, retrieval, and attitude control.

The first mission, TSS–1, will be the upward deployment of a satellite on a 20 km conductive tether. TSS–1 consists of a deployer, a satellite payload of four science instrumentation packages, and four science instrumentation packages mounted on a deployer pallet.

Enabling Technology Development: None

Points Of Contact:

Program Manager:	Rick Howard
Program Scientist:	Dave Evans
NASA Center:	MSFC
Project Manager:	John Price
Project Scientist:	Noble Stone

Section 4 Approved Missions

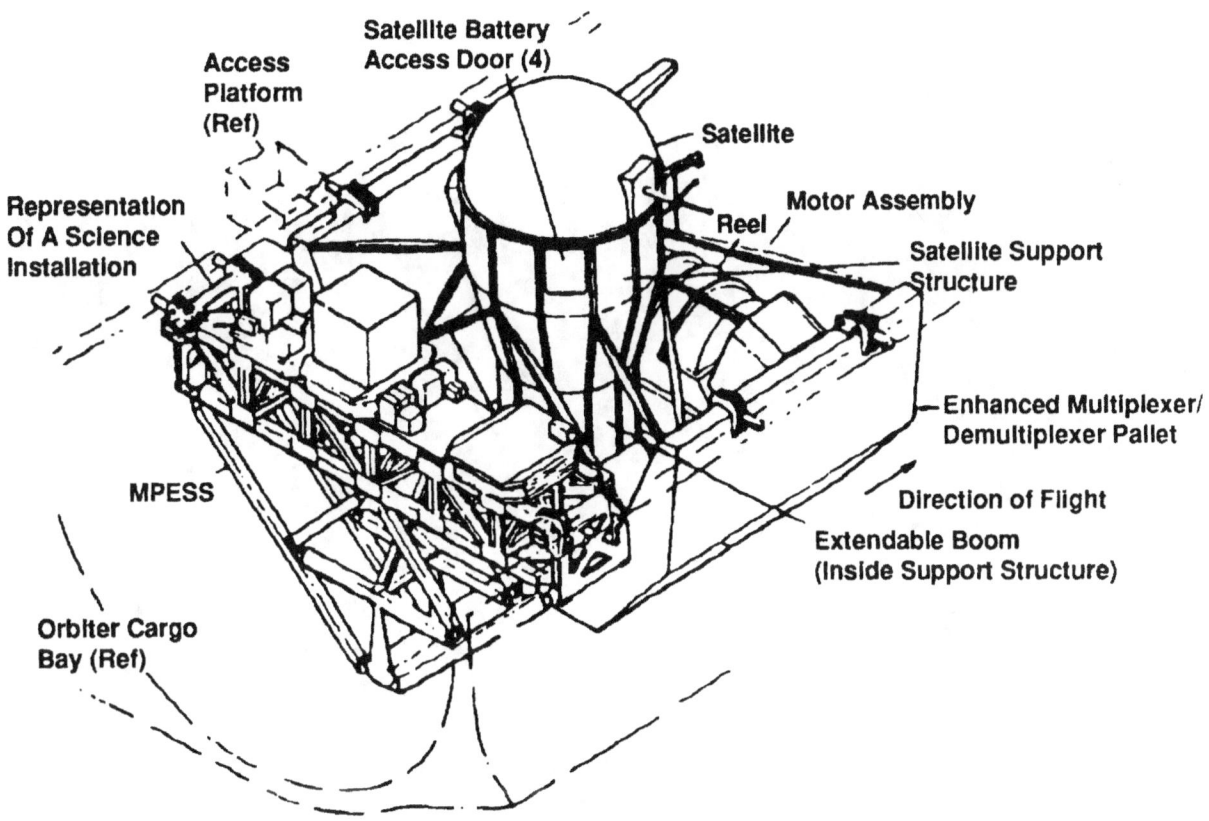

TSS-1 configuration on orbiter

4.4 Wind

Target: Interplanetary medium/solar wind bow-shock region

Orbit:
   a. Lunar swingby (12 Re x 240 Re) on day side
   b. Halo orbit at Sun/Earth L_1 point

Mission Duration: 3 years

Mission Class: Moderate

Mass: 1250 kg

Launch Vehicle: Delta-2 7925 (December 1992)

Theme: Magnetospheric physics

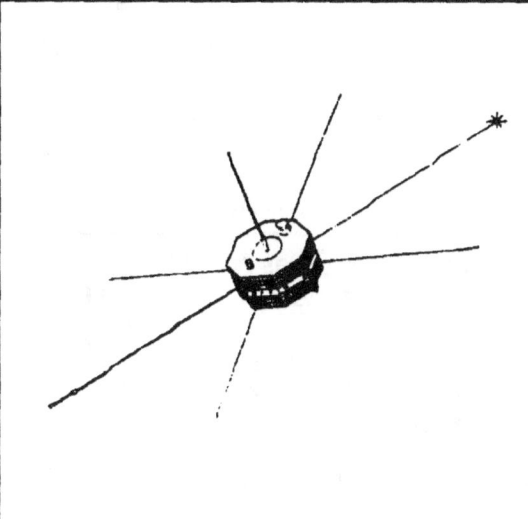

Science Objectives:

The objectives are as follows:

1. Determine the solar wind input properties for magnetospheric and ionospheric studies as part of the ISTP program.
2. Determine the magnetospheric output.
3. Investigate basic plasma processes in the solar wind.
4. Provide baseline observations for global heliospheric studies.

The Wind satellite will specifically measure and investigate:

- Magnetic fields
- Radio and plasma waves
- Hot plasma composition
- Energetic particles
- Solar wind plasma
- Cosmic gamma rays

Spacecraft:

Type: Unique GE-designed observatory, spin stabilized (20 rpm)

Special Features:
- Four 2.2 N spin rate control thrusters
- Eight 22 N spin axis precession thrusters
- 300 kg hydrazine
- Sun Sensor (SSA), Horizon Sensor (HSA), Star Scanner (SSCA)

Special Requirements:
- Spin axis normal to the ecliptic ±1°
- Maximum eclipse duration 90 mins
- Solar radio exclusion zone 3° radius

Instruments:

Investigation	Expt code	Mass (kg)	Data rate (bps)	Principal Investigator
Energetic Particles and 3-D Plasma Analyzer	3-D PLASMA	15.5	1100	R. Lin (Univ. of Calif. Berkeley)
Energetic Particles Acceleration Composition Transport	EPACT	25.9	500	T. Von Rosenvinge (NASA/GSFC)
Magnetic Fields Investigation	MFI	3.3	512	R. Lepping (NASA/GSFC)
Solar Wind Experiment	SWE	12.0	600	K. Ogilvie (NASA/GSFC)
Radio/Plasma Wave Experiment	WAVES	32.0	1000	J. Bouqueret (Meudon Observatory)
Transient Gamma Ray Spectrometer	TGRS	16.6	400	B. Teegarden (NASA/GSFC)
Solar Mass Sensor	SMS	22.9	700	G. Gloeckler (Univ. of Maryland)

Mission Strategy:

Wind may be placed into two different types of orbits: a dayside double-lunar swingby orbit, and a halo orbit around the sunward libration point (L_1).

The spacecraft will be launched into the double-lunar swingby orbit. In this orbit, the line of apsides is held close to the Earth-Sun line throughout the year by means of lunar swingby maneuvers. Data obtained during these passes will be useful for the solar wind input function in support of space physics missions.

After the lunar swingby orbit phase, Wind may be inserted into a small L_1 halo orbit. In this position it will provide optimum interplanetary measurements on a continuous basis.

Enabling Technology Development: None

Points Of Contact:

Program Manager:	Mike Calabrese
Program Scientist:	Tom Armstrong
NASA Center:	GSFC
Project Manager:	Ken Sizemore
Project Scientist:	Keith Ogilvie

Section 4 Approved Missions

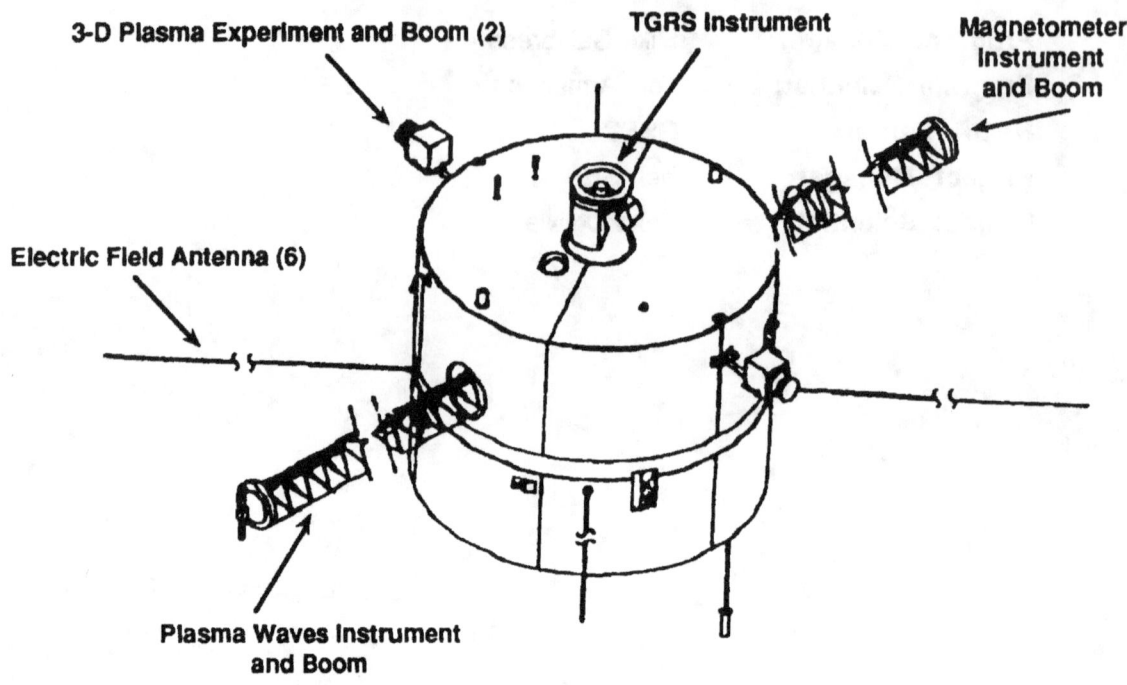

Concept of Wind spacecraft configuration

Section 4 Approved Missions

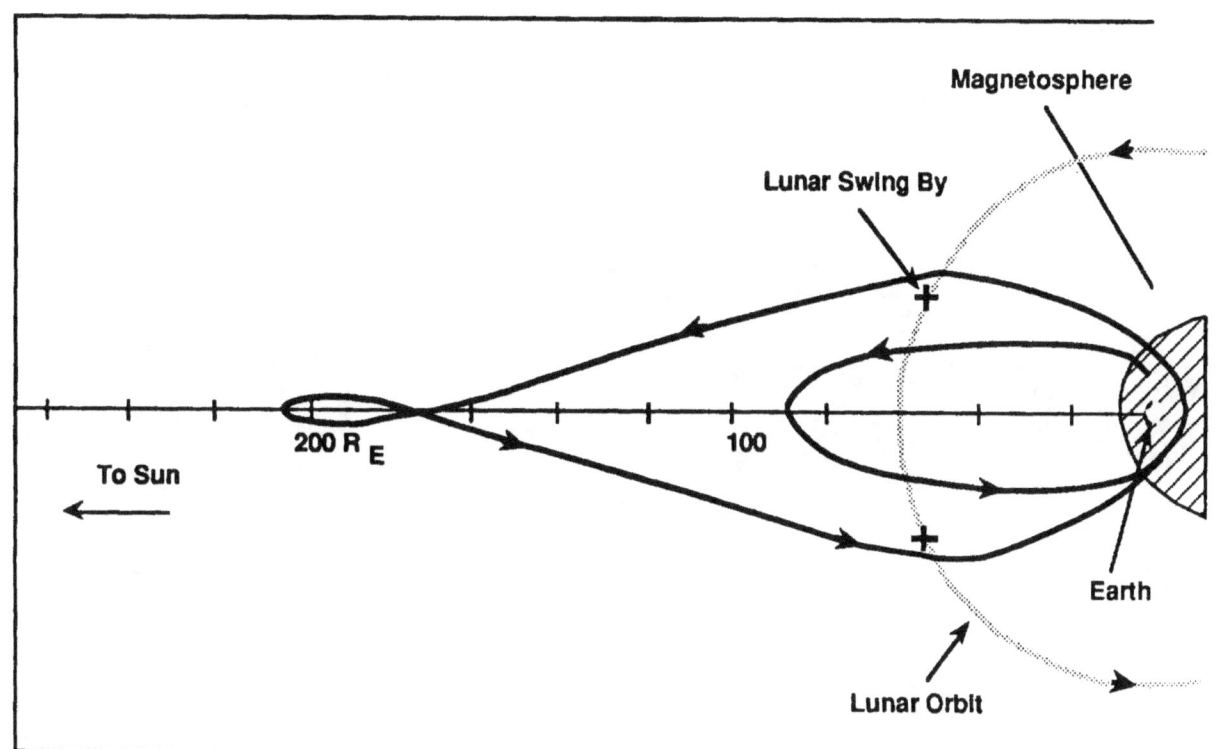

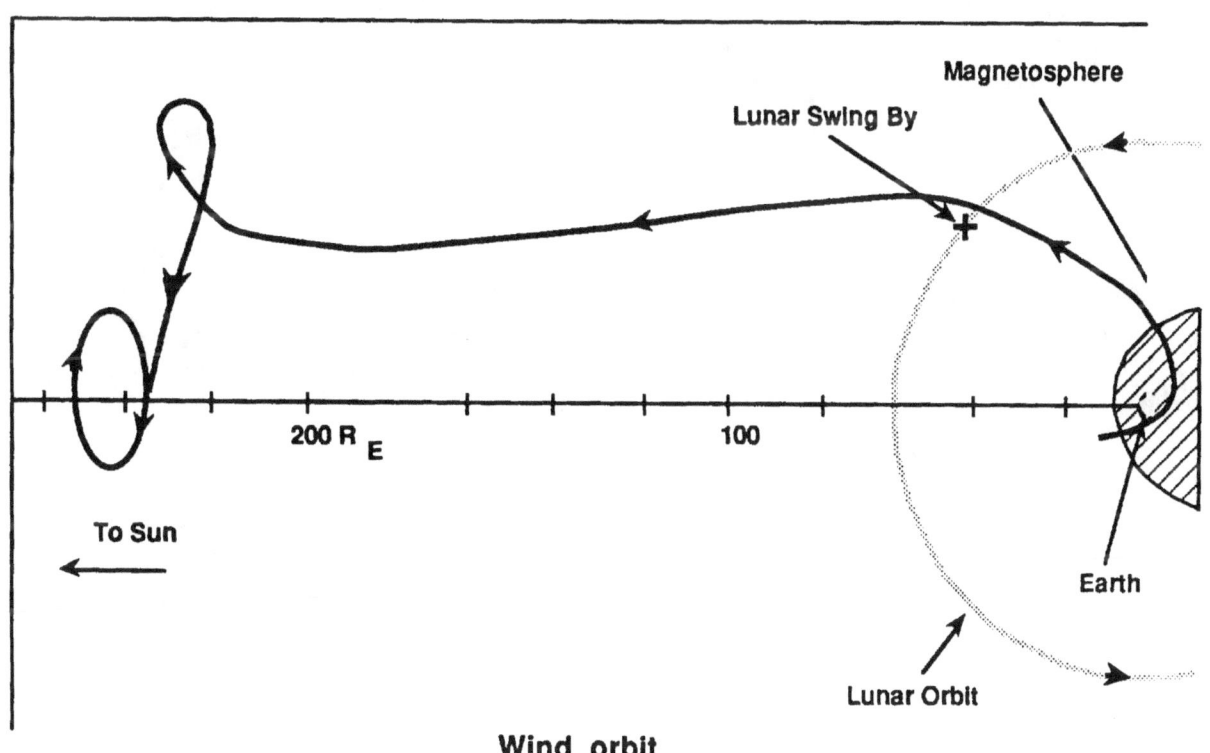

Wind orbit

4.4-5

4.5 Geotail

Target: Earth's magnetotail

Orbit: Distant tail: 220 Re x 8 Re
Near tail: 32 Re x 8 Re

Mission Duration: 3 years

Mission Class: Moderate

Mass: 980 kg

Launch Vehicle: Delta-II (July 1992)

Theme: Magnetospheric physics

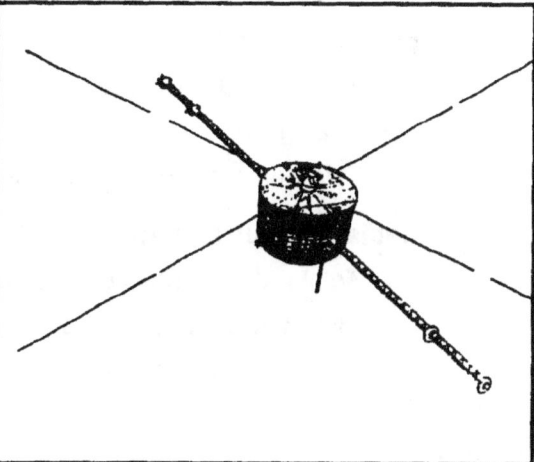

Science Objectives:

Geotail is a collaborative project being undertaken by ISAS and NASA as part of the Collaborative Solar-Terrestrial Research (COSTR) program. The primary objective is to study the dynamics of the Earth's magnetotail over a wide range of distances extending from the near-Earth region (8 Re) to the distant tail (220 Re).

In support of the GGS program, the mission will address transport, storage, and conversion of energy in the tail with a comprehensive instrumentation package for the following:

- Determine the overall plasma electric and magnetic field characteristics of the distant and near geomagnetic tail.

- Help determine the role of the distant and near-Earth tail in substorm phenomena and in the overall magnetospheric energy balance and relate these phenomena to external triggering mechanisms.

- Study the processes that initiate reconnection in the near-Earth tail and observe the microscopic nature of the energy conversion mechanism in this reconnection region.

- Determine the composition and change state of plasma in the geomagnetic tail at various energies during quiet and dynamic periods and distinguish between the ionosphere and solar wind as sources of this plasma.

- Study plasma entry, energization, and transport processes in interaction regions such as the inner edge of the plasma sheet, the magnetopause, and the bow shock; and investigate associated boundary layer regions.

Spacecraft:

Type: Unique ISAS designed observatory, spin stabilized (20 rpm)

Special Features:
- Four 23 N axial thrusters, four 23 N radial thrusters
- 332 kg hydrazine
- Two Sun sensors, one star scanner, one steerable horizon crossing sensor

Special Requirements:
- Total radiation dose should be less than 100 k rads
- Maximum eclipse duration 2 hours
- Solar radio exclusion zone 3°

Instruments:

NASA Element:
- Comprehensive Plasma Experiment (CPI)
- Energetic Particles and Ion Composition (EPIC)
- Multi-Channel Analyzer (MCA)
- Geotail Inboard Magnetometer (GIM)

ISAS Element:
- Electric Field Detector (EFD)
- Magnetic Field Experiment (MGF)
- Low-Energy Particle Experiment (LEP)
- High-Energy Particle Experiment (HEP)
- Plasma Wave Experiment (PWI)

Mission Strategy:

Geotail will spend at least one and a half years in the deep tail, double-lunar-swingby orbit, with a nine-month overlap while Wind, Polar, and CRRES are also in orbit collecting data. After the double-lunar-swingby phase, Geotail will undergo an orbit change to enable observations in the near tail region. During the near tail phase, the spacecraft will be kept in the plasma sheet as long as possible at the time of solstice.

Enabling Technology Development: None

Points Of Contact:

Program Manager:	Mike Calabrese
Program Scientist:	Tom Armstrong
NASA Center:	GSFC
Project Manager:	Ken Sizemore
Project Scientist:	Ron Lepping

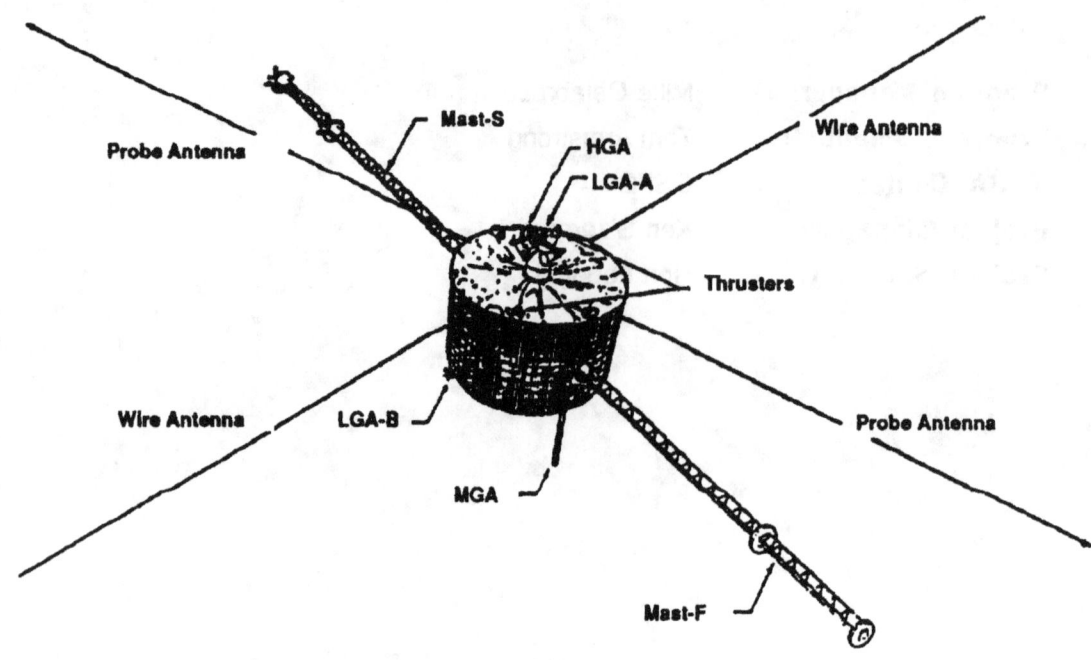

Geotail in-flight configuration

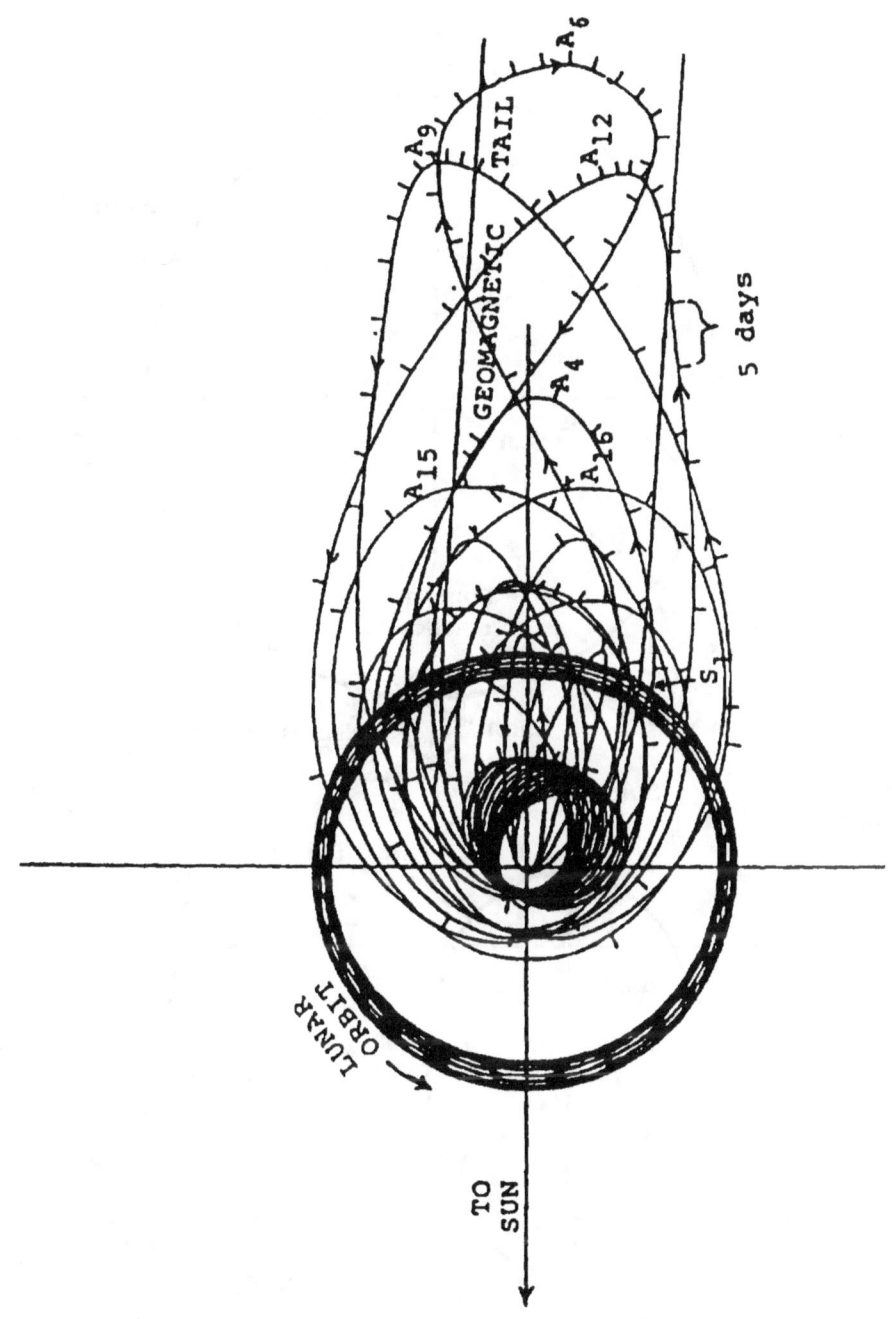

Geotail distant tail orbit (working model)

Section 4 Approved Missions

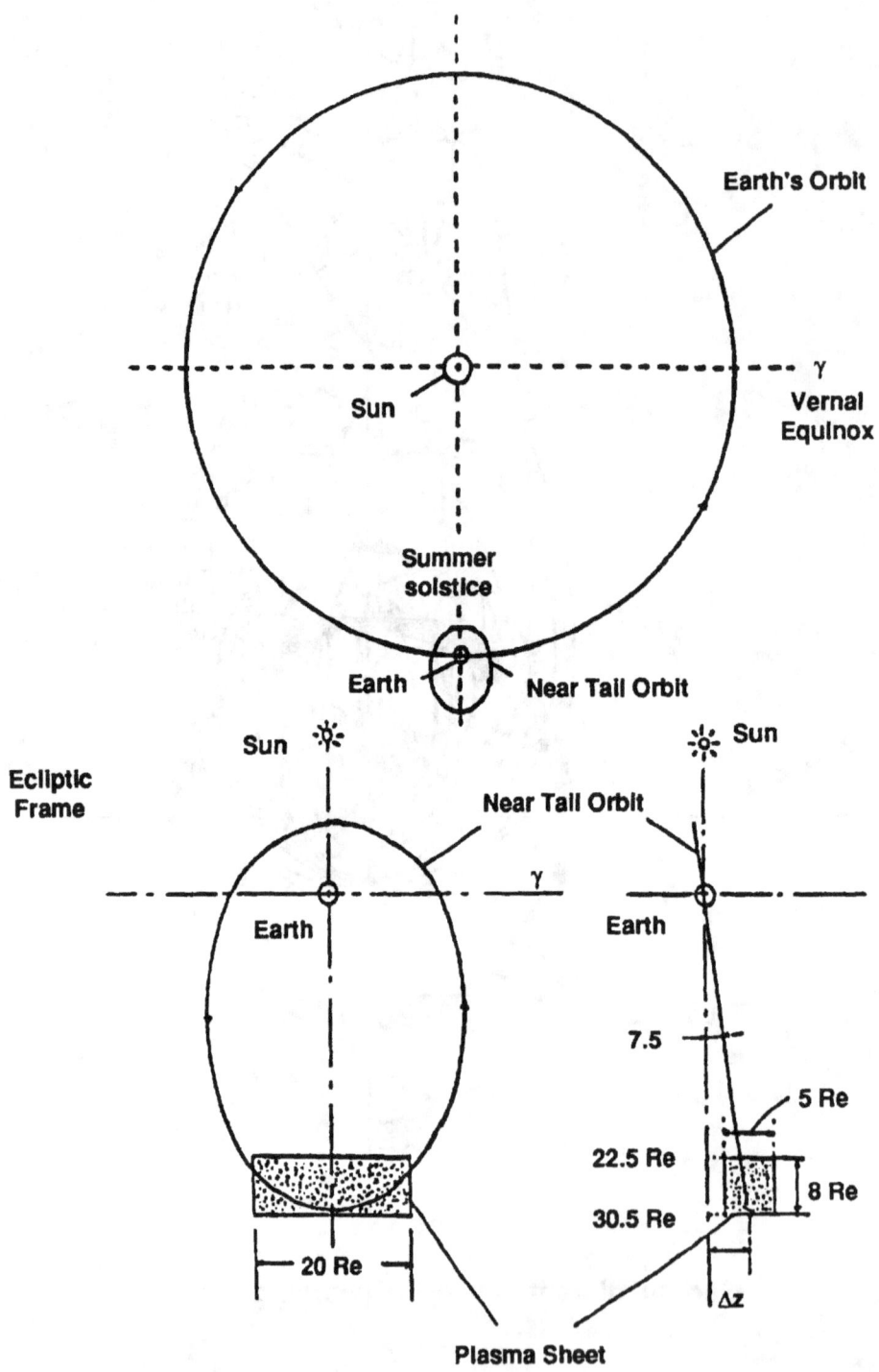

Plasma sheet model and Geotail's near-tail orbit

4.5-6

4.6 Spartan-201

Target: Solar phenomena

Orbit: 160 nmi x 160 nmi

Mission Duration: 40 hours

Mission Class: Small (Spartan)

Mass: 3500 kg

Launch Vehicle: STS

Theme: Solar physics

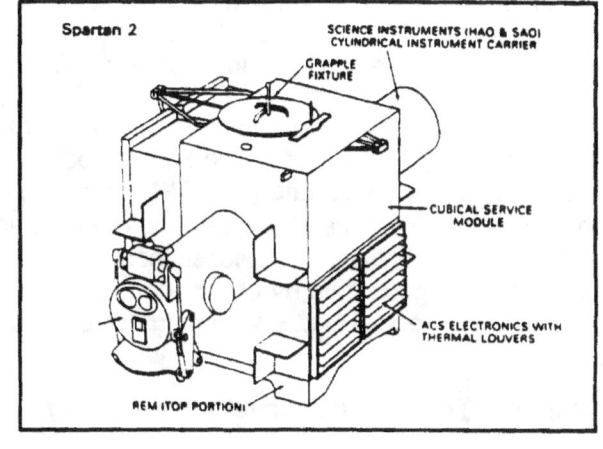

Science Objectives:

There are only 15 minutes of sounding rocket data from a UV coronagraph. Nearly all of it is measurements of the resonantly scattered Lyman alpha. There are only upper limits to O VI intensities and no electron-scattered Lyman alpha data. Thus Spartan data will provide critical information for developing science software and observing programs for UVCS/SOHO. (It is far easier to develop these if the software and observing programs are based on experimental experience in addition to theoretical predictions.) The electron temperature is one of the most fundamental solar wind parameters because electron thermal conduction is one of the primary means of energy transport in the corona and because the electron temperature controls the ion states in the low corona (before they "freeze in" in the solar wind). If you are going to determine the role of thermal pressure gradients in driving the solar wind (and therefore the need for other mechanisms such as wave-particle interactions), it is critical to measure the electron temperature. In short, determination of the electron temperature is critical to understanding the heating and acceleration of the solar wind.

Spacecraft:

   Type: Shuttle subsatellite
   Special Features: TBD
   Special Requirements: TBD

Instruments:

- UV coronagraph

Mission Strategy:

Spartan-201 will be deployed from the shuttle bay and retrieved on the same mission, 40 hours later.

Spartan-201 is the next step toward understanding the heating and acceleration of the solar wind. After 15 minutes of rocket flight, in 1982 Spartan-201 was slated to fly in 1985 to acquire longer duration observations for determining electron temperature and flow speeds of higher stages of ionization. Weak photon fluxes preclude rocket measurements of these quantities. Several launch slips occurred (forced by Halley's comet plus Challenger) and Spartan-201's most recent manifested flight in December 1991 has now been pushed out to 1993.

Spartan-201 serves as a prototype for important aspects of the Ultraviolet Coronagraph-Spectrometer (UVCS) and Large-Field Spectrographic Coronagraph (LASCO) on the Solar and Heliospheric Observatory (SOHO). Spartan-201 represents the only source of data on solar wind generation and coronal heating with which to focus the international SOHO science preparations. (The SOHO helioseismology experiments, on the other hand, have been supplied analogous data by observatories on the ground.) Because the Spartan-201 data are unique, they are vital in strengthening our ability to deal with SOHO data.

Enabling Technology Development: None

Points Of Contact:

Program Manager:	Glenn Mucklow
Program Scientist:	Dave Bohlin
NASA Center:	GSFC
Project Manager:	Frank Collins
Project Scientist:	TBD

Section 4 Approved Missions

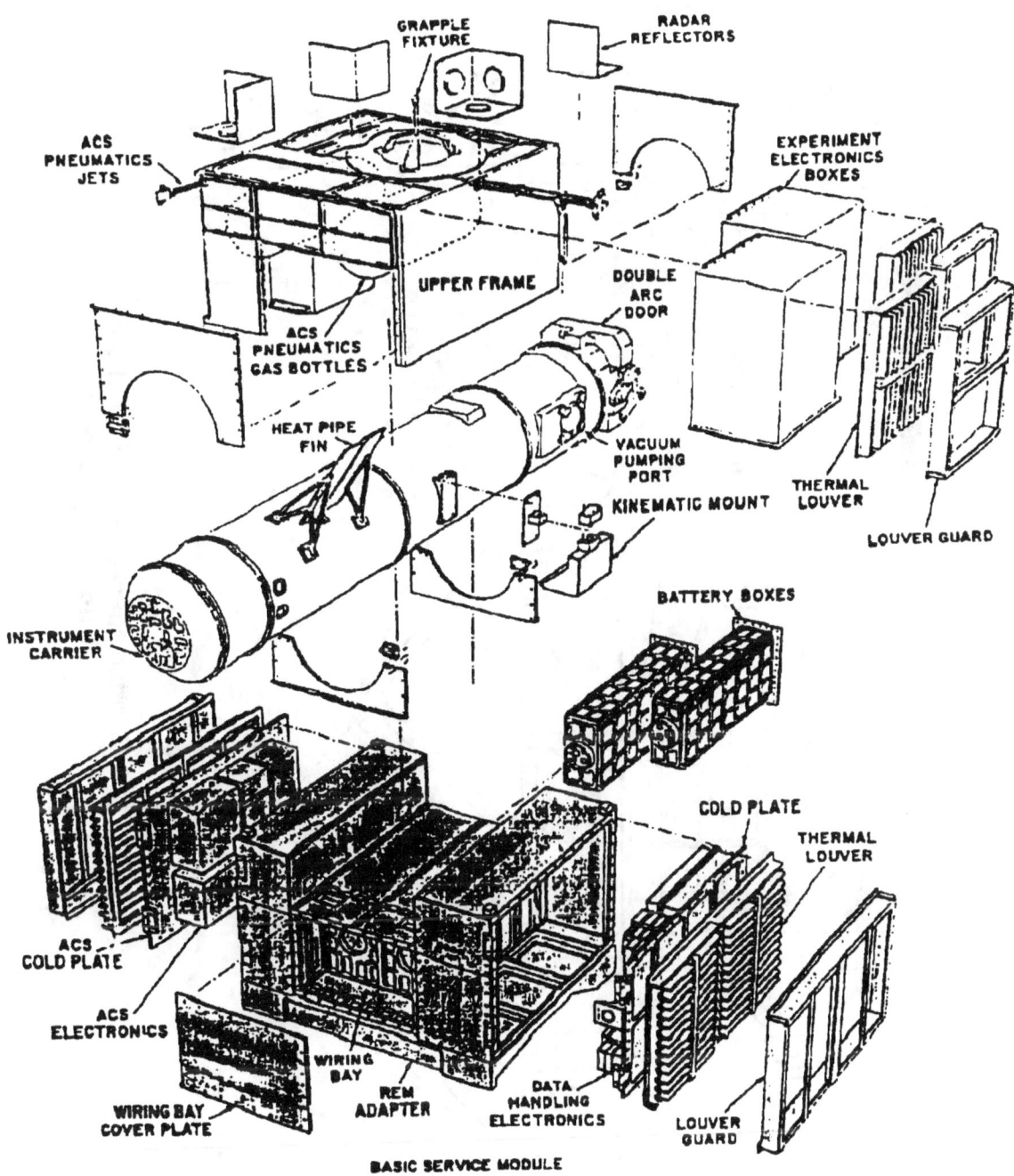

Spartan-201

Section 4 Approved Missions

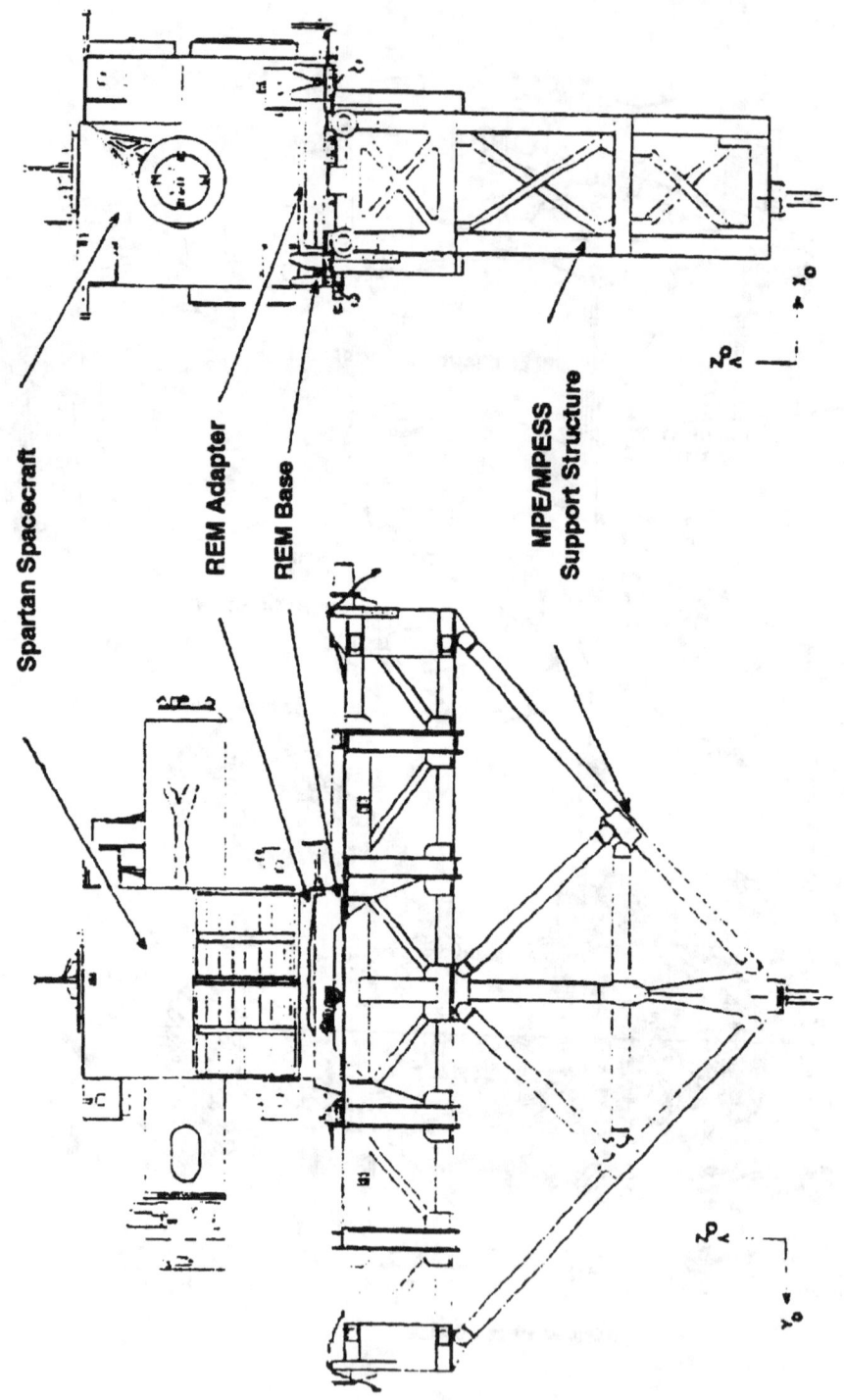

Spartan spacecraft configuration: all up configuration

Section 4 Approved Missions

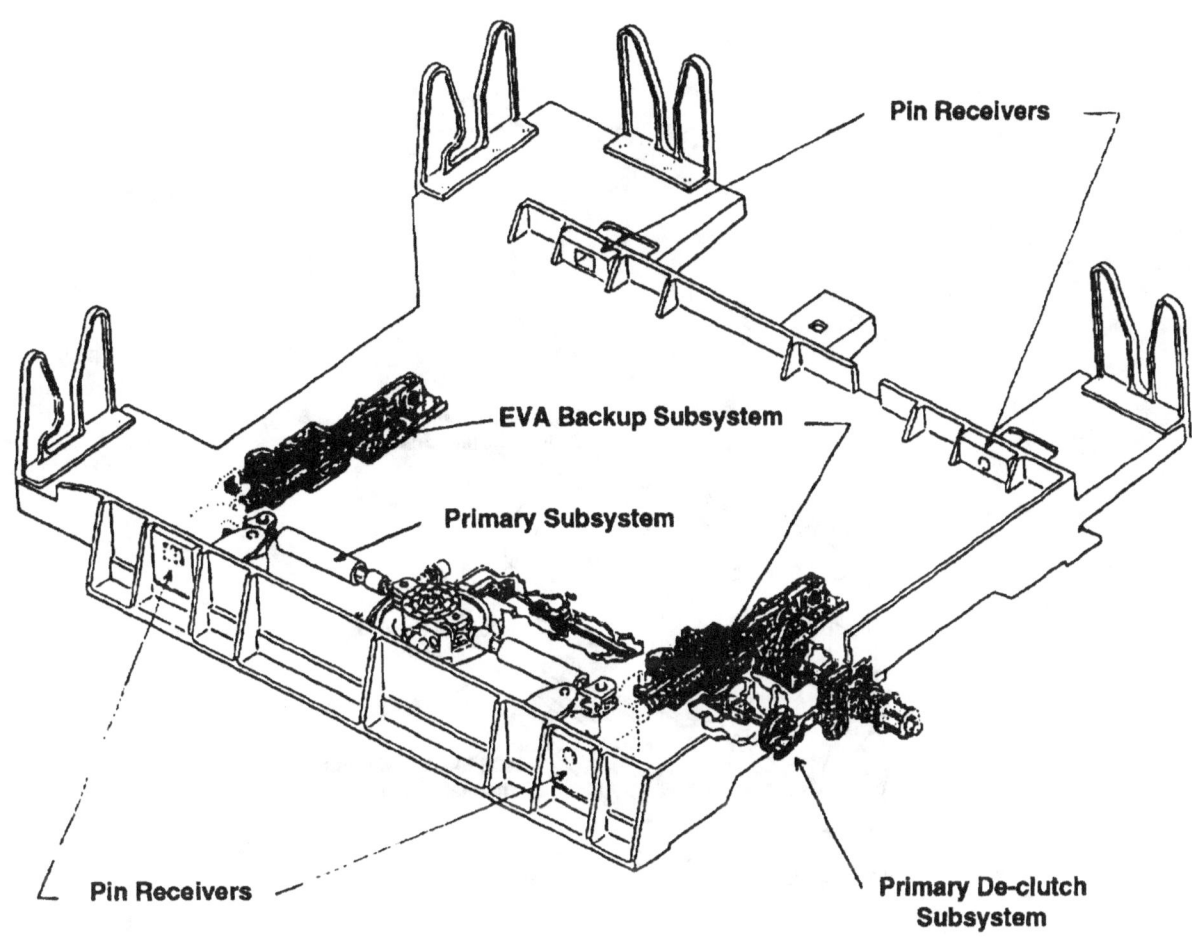

**Spartan Release Engine Mechanism (REM):
REM base and subsystems**

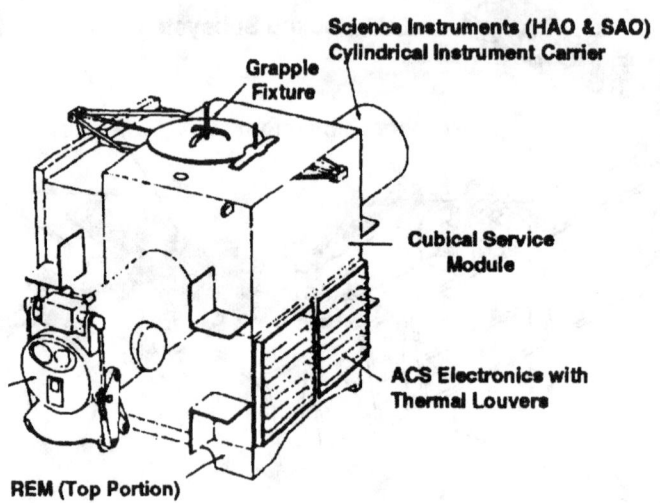

Spartan 2

4.7 Waves in Space Plasma (WISP)

Target: Space plasma in Earth's ionosphere

Orbit: 60 nmi x 160 nmi, high inclination desired

Mission Duration: 7 days

Mission Class: Small

Mass: 5300 kg

Launch Vehicle: STS

Theme: Ionospheric physics

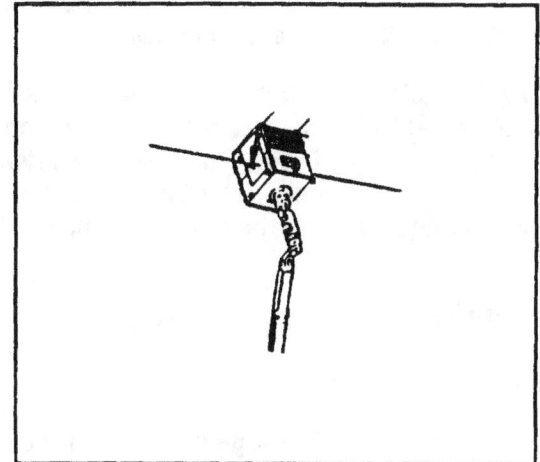

Science Objectives:

WISP will study wave-particle interactions, determine propagation paths, map the region of near-Earth space accessible for communications, and remotely sense variations in electron density.

The objective of the Waves in Space Plasma (WISP) investigation is to conduct a series of active experiments using a high-frequency radio transmitter to perturb the space plasma environment while simultaneously making *in situ* and remote measurements of the response of the environment to such controlled perturbations.

Spacecraft:

Type: Shuttle/Spartan
Special Features: TBD
Special Requirements: TBD

Instruments:

Mission Strategy:

WISP will use a high-frequency radio transmitter on board the shuttle orbiter to sound the space plasma environment. Simultaneously, the Spartan vehicle released from the orbiter bay will carry sensors to make *in situ* and remote measurements of the response of the environment to such controlled soundings.

Transmitters, antennae, and diagnostic instruments will be mounted on an Enhanced Multiplexer Pallet (EMP) which is in the shuttle bay. Receiver electronics and antennae will be attached to Spartan. During the periods when the Spartan is deployed from the shuttle, plasma

Section 4 Approved Missions

waves will be generated in a controlled manner from the orbiter at radio frequencies up to 30 MHz and measured both on board the shuttle and on the Spartan vehicle. In this manner, the WISP will study wave-particle interactions, determine propagation paths, map the region of near-Earth space accessible for communications at these radio frequencies, and remotely sense variations in electron densities.

The Canadian-provided Control Electronics Unit, Radio Frequency Unit, Power Interface Unit, Transmitter Unit, Antenna Interface Unit, and Computer (*) along with the U.S.-provided 30 to 50-meter tip-to-tip Dipole Antenna Subsystem (DASS) will be mounted on the EMP which is attached to the shuttle. The Canadian-provided Power Distribution Module, Synthesizer Module, and Data Interface Subsystem(*) will be attached to the OMV.

Enabling Technology Development: None

Points Of Contact:

Program Manager:	Bill Piotrowski/Glenn Mucklow (Spartan)
Program Scientist:	Dave Evans
NASA Center:	MSFC
Project Manager:	TBD
Project Scientist:	TBD

Section 4 Approved Missions

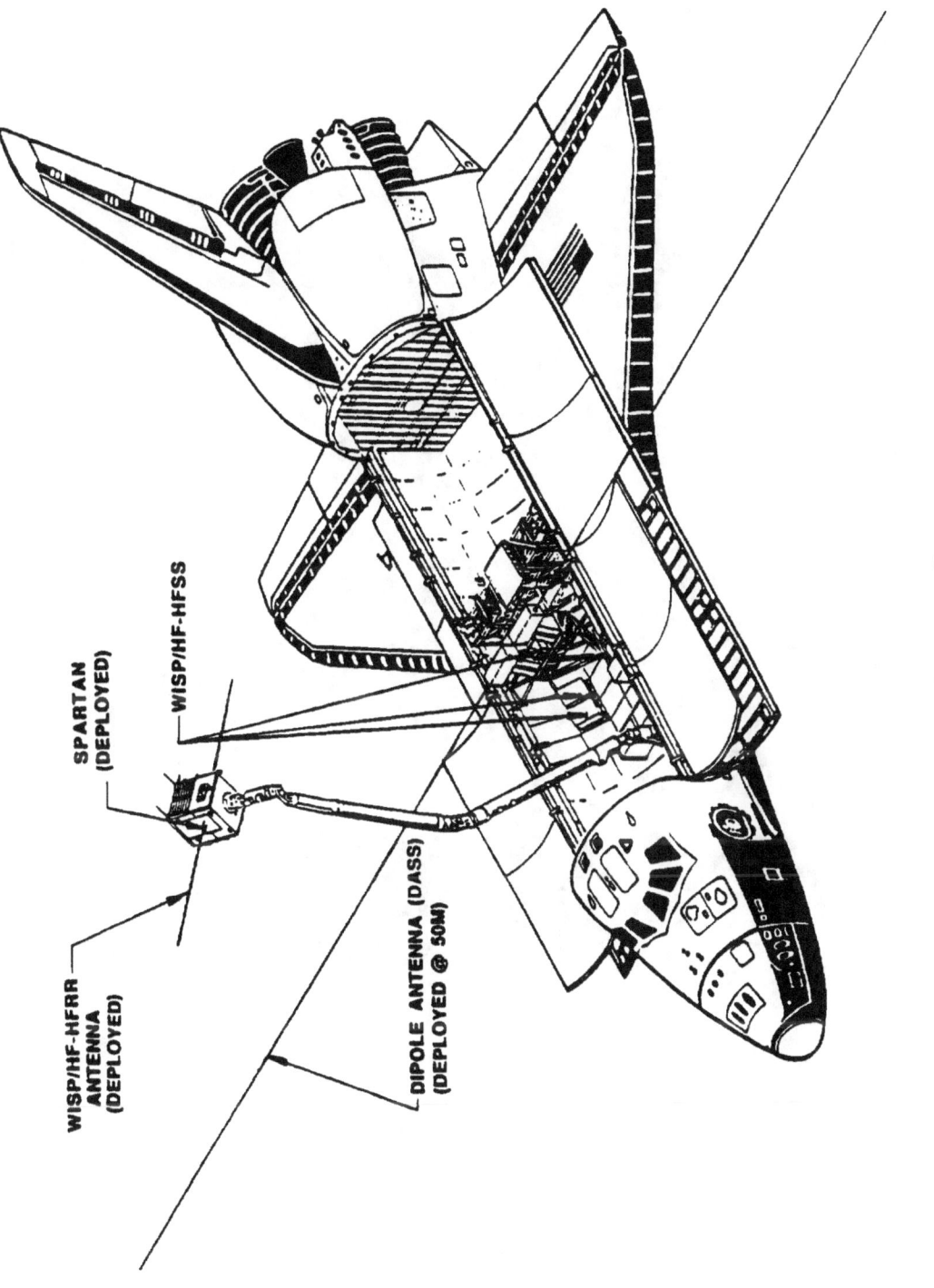

WISP

4.8 Polar

Target: Magnetosphere and ionosphere

Orbit: 1.8 Re x 9 Re, polar orbit

Mission Duration: 3 years nominal

Mission Class: Medium

Mass: 1250 kg

Launch Vehicle: Delta-2 (June 1993)

Theme: Ionospheric physics and polar magnetosphere

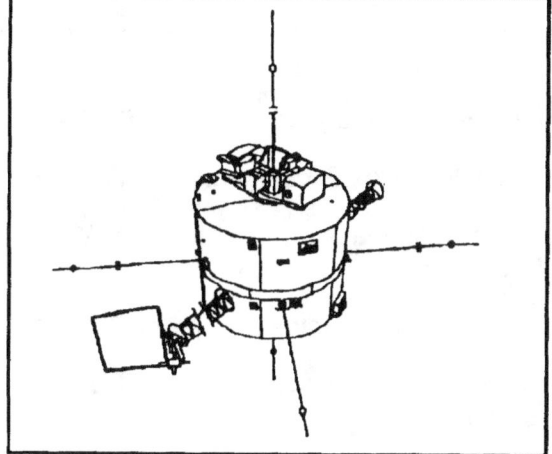

Science Objectives:

Polar will be the second-launched of two NASA spacecraft in the GGS initiative, which is part of the ISTP project. The objectives of the Polar mission are to:

- Characterize the energy input to the ionosphere.
- Determine role of ionosphere in substorm phenomena.
- Measure complete plasma, energetic particles, and fields in the polar regions.
- Determine characteristics of ionospheric plasma outflow.
- Provide global multispectral auroral images of the footprint of the magnetospheric energy disposition into the ionosphere and upper atmosphere.

Spacecraft:

Type: Unique GE-designed observatory, spin stabilized (10 rpm)

Special Features:
- Four 2.2 N spin-rate thrusters, four 4.4 N spin-axis precession thrusters, four 22 N axial velocity change thrusters
- 338 kg hydrazine
- Sun sensor, horizon sensor

Special Requirements:
- Maximum eclipse duration 90 mins
- Antenna pointing to within 1°

Instruments:

Investigation	Expt code	Mass (kg)	Data rate (bps)	Principal Investigator
Plasma Wave Instrument	PWI	18.4	2520	D. Gurnett (Univ. of Iowa)
Fast Plasma Analyzer	HYDRA	14.4	4400	J. Scudder (GSFC)
Magnetic Fields Experiment	MFE	5.0	500	C. Russell (UCLA)
Toroidal Ion Mass Spectrograph	TIMAS	16.5	3600	E. Shelley (LPARL)
Electric Fields Instrument	EFI	31.9	2500	F. Mozer (UCB)
Thermal Ion Dynamics Experiment	TIDE	33.3	2520	C. Chappell (MSFC)
Ultraviolet Imager	UVI	18.0	12000	M. Torr (MSFC)
Visible Imaging System	VIS	24.0	11000	L. Frank (Univ. of Iowa)
Polar Ionospheric X-Ray Imaging Experiment	PIXIE	24.5	3500	W. Imhoff (LPARL)
Charge and Mass Magnetospheric Ion Composition Experiment	CAMMICE	12.9	1280	T. Fritz (LANL)
Comprehensive Energetic Particle Pitch Angle Distribution	CEPPAD	14.4	4380	B. Blake (Aero)

Mission Strategy:

In support of the GGS program, with Wind, Geotail, and CRRES overlapping, Polar will be placed in a 90° inclination orbit with an apogee of 8–10 Re and a perigee of 1.8 Re. The perigee radius will be increased to the maximum allowed by the spacecraft final mass and capability of the on board propulsion system. This will maximize coverage of the low-altitude auroral acceleration region. The apogee may be lowered several years after launch. In this orbit, Polar will provide coverage of the dayside cusps region at high latitudes and the southern hemisphere polar cusps at low altitudes, as well as global imaging of the northern auroral zone.

Enabling Technology Development: Pointing platform

Points of Contact:

Program Manager:	Mike Calabrese
Program Scientist:	Tom Armstrong
NASA Center:	GSFC
Project Manager:	Ken Sizemore
Project Scientist:	Jack Scudder

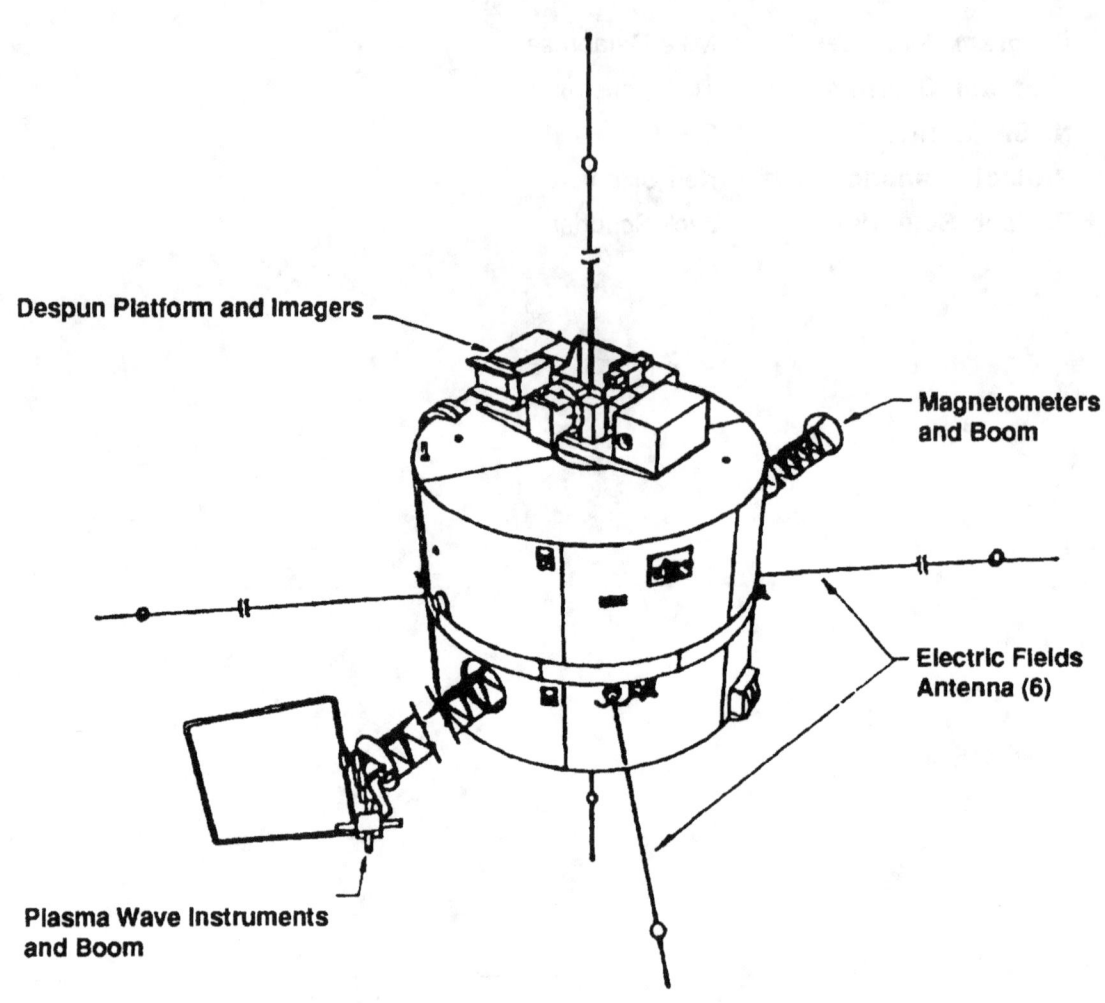

Concept of Polar spacecraft configuration

Section 4 Approved Missions

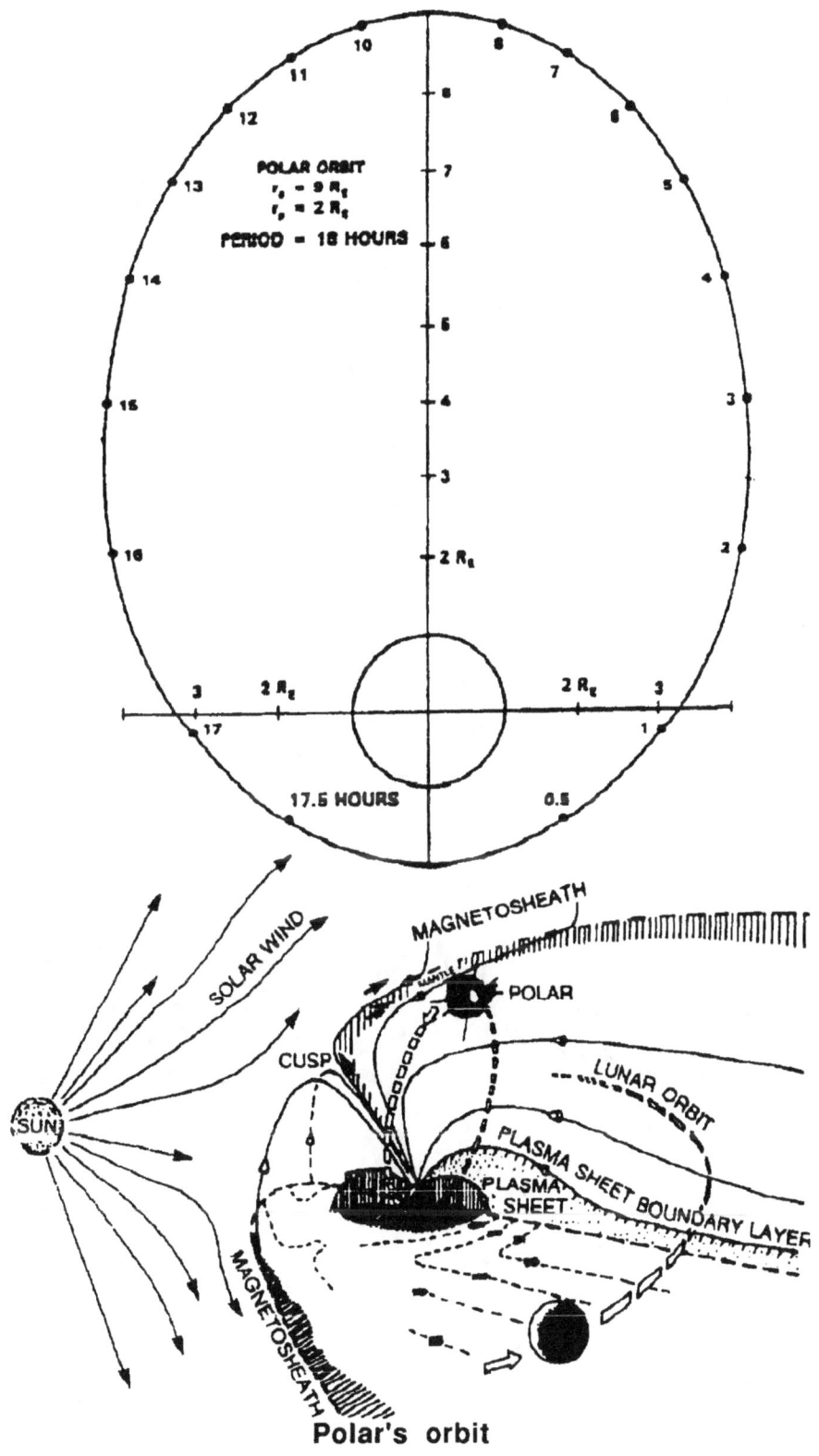

Polar's orbit

4.9 Cluster

Target: Earth's magnetosphere and solar wind

Orbit: Polar; 3 Re x 20 Re elliptic

Mission Duration: 2 years

Mission Class: Moderate

Mass: 4000 kg

Launch Vehicle: Ariane 5 (December 1995)

Theme: Magnetospheric physics

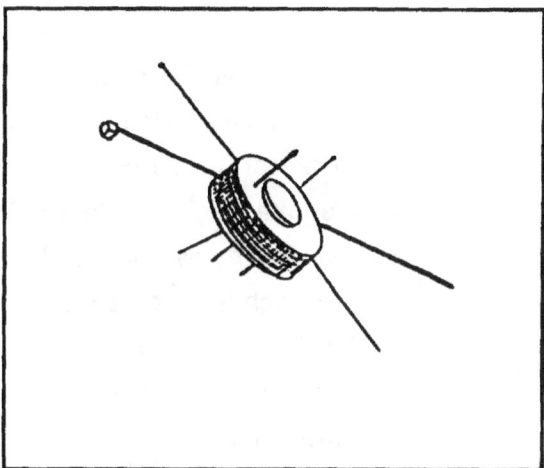

Science Objectives:

The objective of this mission is to perform three-dimensional studies of the microphysical properties of different plasma states in the Earth's magnetosphere and solar wind. The Cluster satellite will specifically investigate the following:

- Magnetic fields
- Electric fields
- Plasma wave
- Plasma RF sounder
- Energetic ion and electron distribution
- Hot plasma ion and electron distribution

Spacecraft:

Type: Unique ESA-designed observatory, spin stabilized (15 rpm)

Special Features:
- MMH + MON bi-propellant 400 N injection thrusters
- 640 kg fuel

Special Requirements:
- Spacecraft to be maintained in controlled tetrahedron configuration while in orbit
- Orbit to be stable over 2 years

Instruments:

- Active Spacecraft Potential Control Experiment (ASPOC)
- Cluster Ion Spectrometry (CIS)
- Digital Wave Processor (DWP)
- Electron Drift Instrument (EDI)
- Spherical Double Probe Electrical Field Experiment (EFW)
- Fluxgate Magnetometer (FGM)
- Plasma Energy Angle and Composition Experiment (PEACE)
- Research & Adaptive Particle Imaging Detector (RAPID)
- Spatial Temporal Analysis of Field Fluctuations (STAFF)
- Wideband Plasma Wave & Radio Interferometry (WBD)
- Waves of High Frequency and Sounder for Probing of Electron Density by Relaxation (WHISPER)

Mission Strategy:

Cluster, a joint NASA/ESA venture, will consist of four identically-instrumented spacecraft launched by ESA into a 3 Re x 20 Re elliptical polar orbit as part of the STSP and ISTP programs. The orbit will pass through the northern cusp region, and cross the day side magnetopause at lower latitudes. The apogee should be large enough to cross the bowshock, and the fully developed magnetotail should be crossed at distances larger than 12 Re.

The four spacecraft will be injected into a standard GTO, then transferred into a polar orbit via single lunar flyby, double lunar flyby, or direct injection selected such as to allow the exploration of the bowshock, the magnetopause, the dayside cusp, and the geomagnetic tail current sheet.

Enabling Technology Development: None

Points of Contact:

Program Manager:	Mike Calabrese
Program Scientist:	Tom Armstrong
NASA Center:	GSFC
Project Manager:	Ken Sizemore
Project Scientist:	Mel Goldstein

Section 4 Approved Missions

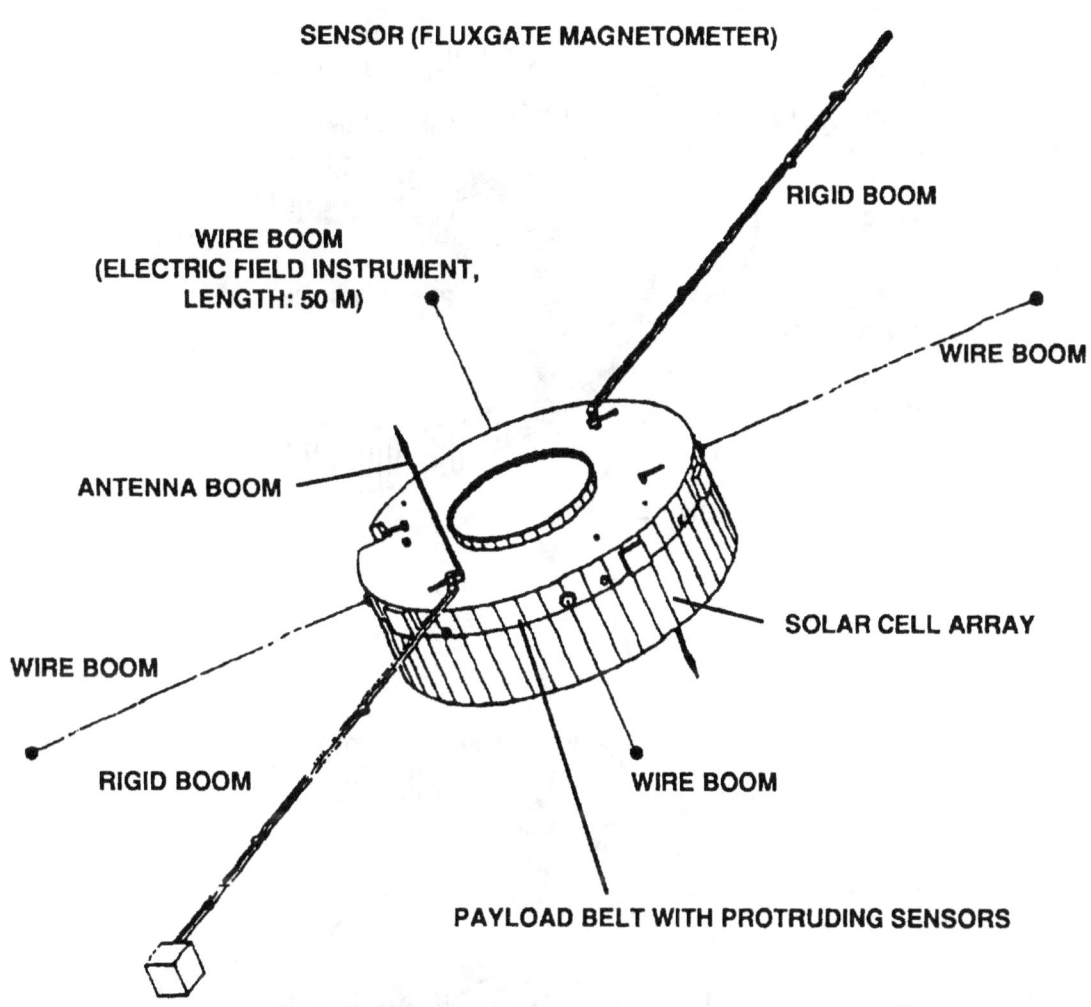

One of the four identical Cluster spacecraft in flight configuration as defined during the Phase A Study

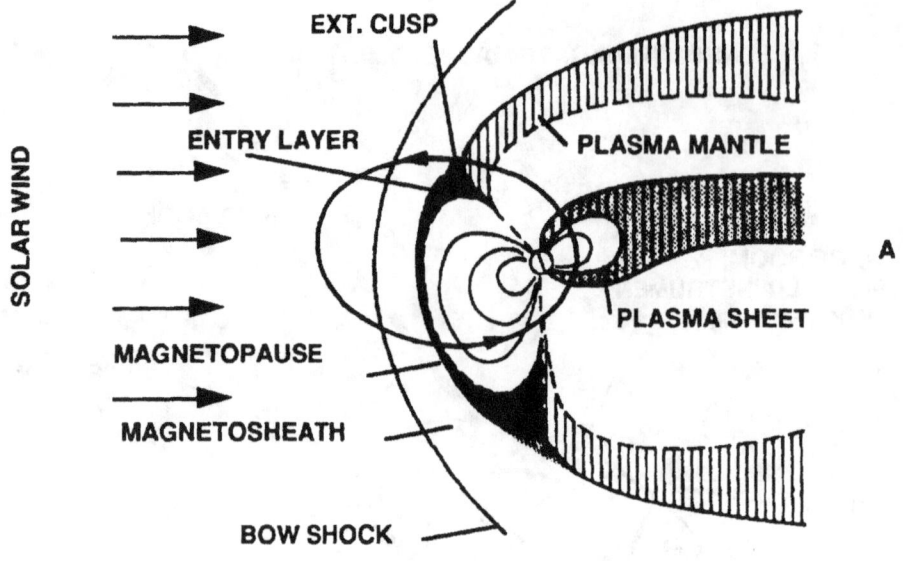

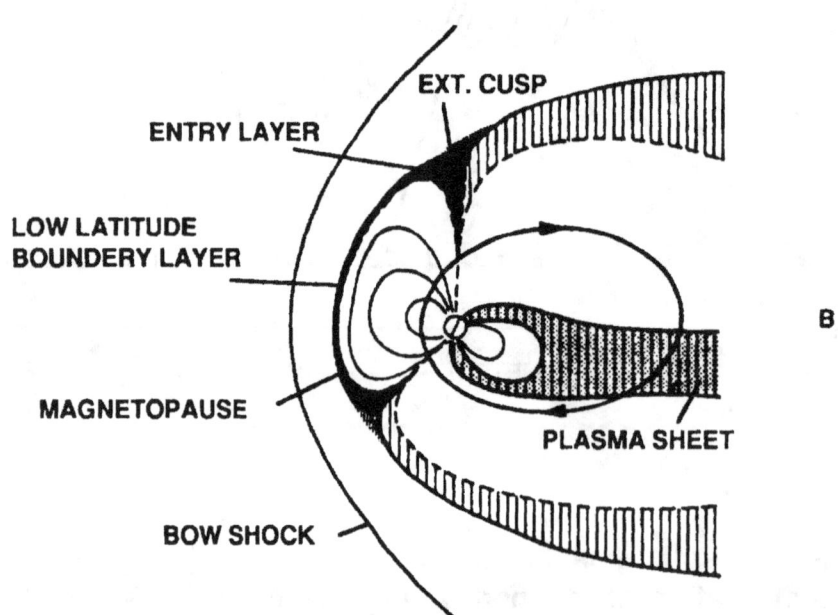

Cluster orbit at six month intervals in relation to the magnetosphere, dayside (a) and nightside (b) orbit

4.10 Solar and Heliospheric Observatory (SOHO)

Target: Solar corona

Orbit: Heliocentric; halo around Sun-Earth L$_1$ Lagrangian point
- Amplitude (x) 2000,000 km
- Amplitude (y) 6500,000 km
- Amplitude (z) 12000,000 km
- Period 6 months

Mission Duration: 2 years baseline, 6 years possible

Mission Class: Moderate–major

Mass: 1800 kg (dry)

Launch Vehicle: Atlas II (March 1995)

Theme: Solar physics

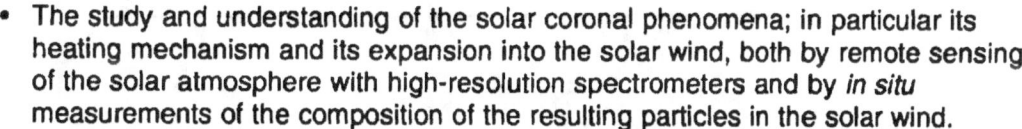

Science Objectives:

The objectives of the SOHO mission are:

- The study and understanding of the solar coronal phenomena; in particular its heating mechanism and its expansion into the solar wind, both by remote sensing of the solar atmosphere with high-resolution spectrometers and by *in situ* measurements of the composition of the resulting particles in the solar wind.

- The study of the solar structure and interior dynamics from its core to the photosphere by helioseismological methods, and the measurement of the solar irradiance variations.

To fulfill these objectives, the payload will contain instruments dedicated to: high-resolution coronal observations, *in situ* measurements of the solar wind, solar irradiance, and helioseismology.

Spacecraft:

Type: Unique ESA 3-axis stabilized observatory design

Special Features: TBD

Special Requirements: TBD

Instruments:

	Height (kg)	Power (W)
Coronal Diagnostic Spectrometer (CDS)	8.4	4.5
Michelson Doppler Imager (MDI)	43.4	55.0
Solar Ultraviolet Measurements of Emitted Radiation (SUMER)	88.0	35.0
Charge, Element & Isotope Analysis (CELIAS)	24.5	18.0
Ultraviolet Coronagraph Spectrometer (UVCS)	107.5	35.0
Extreme Ultraviolet Imaging Telescope (EIT)	17.5	27.5
Wide Field White Light & Spectrometric Coronagraph (LASCO)	57.4	41.0
Energetic and Relativistic Nuclear and Electron Experiment (ERNE)	TBD	TBD
Global Oscillations at Low Frequencies (GOLF)	31.2	30.0
A Study of Solar Wind Astrophies (SWAN)	11.6	9.5
Variability of Solar Radiance and Gravity Oscillation (VIRGO)	14.6	16.6
Comprehensive Suprathermal & Energetic Particle Analyzer (COSTEP)	18.5	22.0

Mission Strategy:

SOHO is a joint venture of ESA and NASA within the framework of the Solar Terrestrial Science Program (STSP), and will participate in the ISTP project. The SOHO spacecraft is baselined for a 1995 launch using an Atlas Centaur launch vehicle. From a parking orbit it will be accelerated into a trajectory towards the L_1 Lagrangian point, a position of gravitational zero that exists at a distance of approximately 1.5 million km along the Earth/Sun line. The "cruise" phase to L_1 will take some 4 months, during which time the experiments will remain dormant. Since halo orbits around the L_1 point are highly unstable, regular maneuvers of the spacecraft will be needed to maintain the orbit. The lifetime of the satellite is nominally 2 years, but will be extendable to 6 years.

The concept of controlling the SOHO mission is based upon the use of the Deep Space Network, with the mission and experiment control centers located at the NASA Goddard Space Flight Center (GSFC). Provision will be made at GSFC for the experimenters to integrate their own computers with the main spacecraft data distribution system in an area to be known as the Experiment Operations Facility (EOF).

The size and configuration of the SOHO spacecraft are dominated by the size of the large coronal instruments, and the accommodation of the liquid propulsion system. The instruments will be mounted on a dedicated Payload Module that will allow flexibility during the integration phase of the spacecraft. Spacecraft pointing constraints currently provide for 1 arc second stability over a 15 minute period, with the boresight always pointed at Sun center to match the requirements of the helioseismology and coronagraph instruments (this necessarily means that other experiments, requiring second offset pointing, will have to provide their own pointing system).

Enabling Technology Development: TBD

Points of Contact:

Program Manager:	Mike Calabrese
Program Scientist:	Dave Bohlin
NASA Center:	GSFC
Project Manager:	Ken Sizemore
Project Scientist:	Art Poland

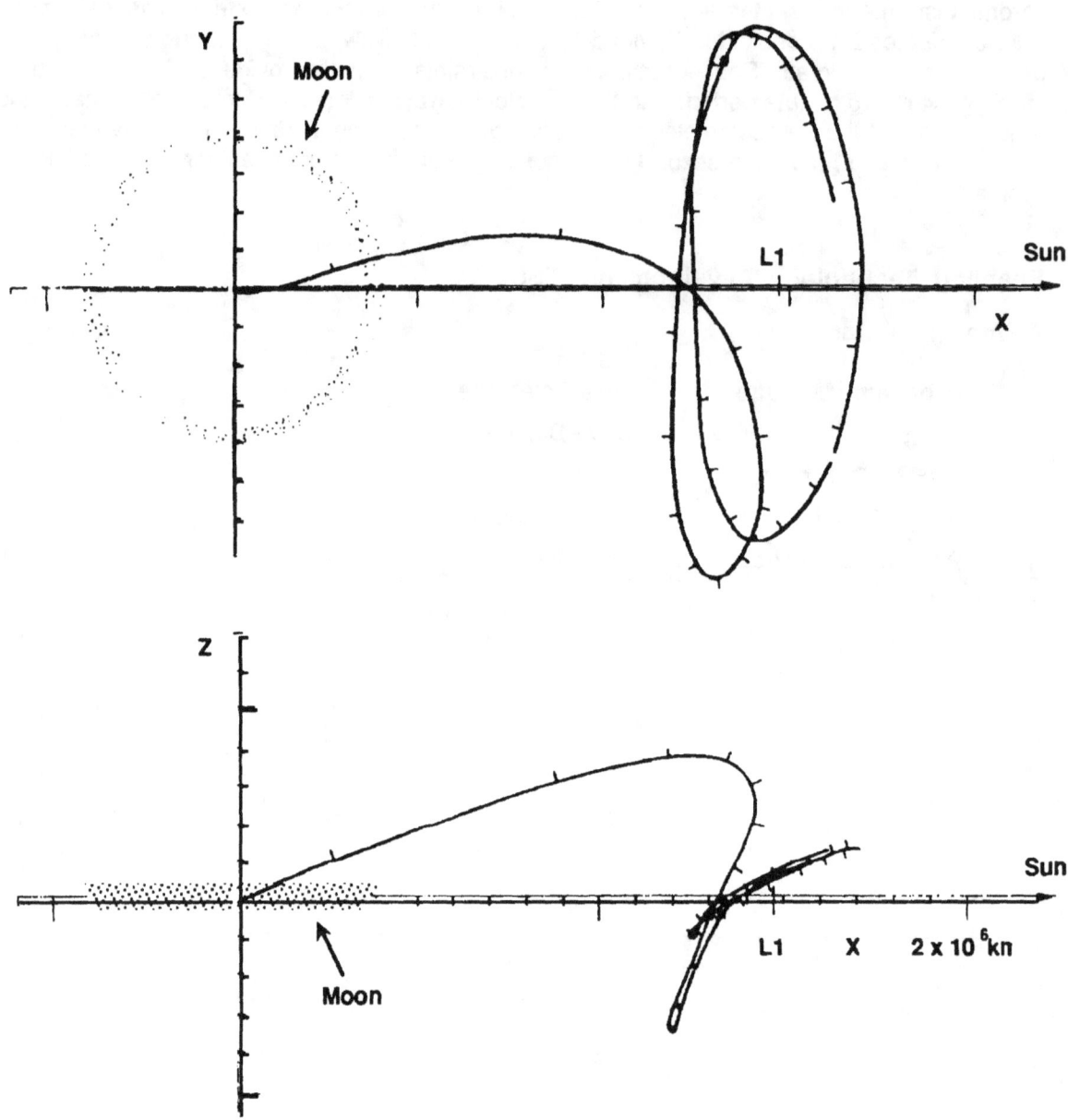

SOHO insertion trajectory and halo orbit represented in geocentric solar ecliptic coordinates. Projection in the ecliptic plane (XY) and in the normal to ecliptic plane (ZX). Tick marks are 10 days apart.

Section 4 Approved Missions

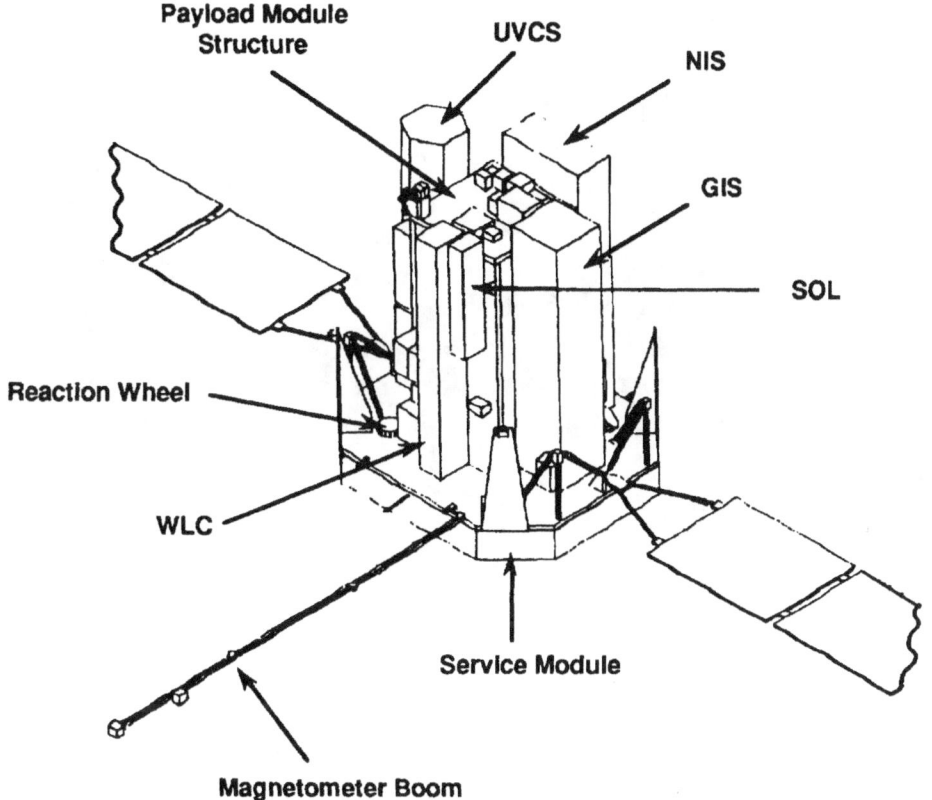

SOHO semi-exploded view

Section 4 Approved Missions

4.11 Energetic Heavy Ion Composition (EHIC)

Target: Low-energy solar flare elemental and isotopic composition

Orbit: 833 km x 833 km, or 870 km x 870 km, at 99° inclination (TIROS)

Mission Duration: 3 years

Mission Class: Secondary payload/ Explorer program

Mass: TBD

Launch Vehicle: Atlas E or Titan II, 1991

Theme: Cosmic and heliospheric physics

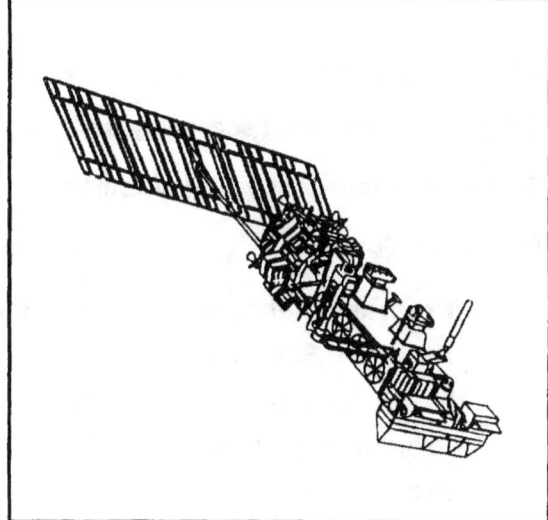

Science Objectives:

- Measure the chemical composition of solar energetic particles from hydrogen to nickel.
- Measure the isotopic composition of solar energetic hydrogen and helium nuclei.
- Study the enrichment of 3He and heavy ions in solar flares.
- Investigate the acceleration mechanisms for nuclei in solar flares.
- Study the transport of energetic particles in the heliosphere and corona.
- Measure the composition and energy spectra of energetic particles in the Earth's magnetosphere.

Spacecraft:

    Type: Advanced TIROS NOAA–I or –J (ATN–I or –J)

    Special Features: TBD

    Special Requirements: TBD

Instruments:

The EHIC instrument is comprised of two separate charged particle telescopes: (1) the Low Energy Particle Analyzer (LEPA), provided by the National Research Council of Canada will cover the very lowest energies, up to about 15 MeV/nucleon for iron nuclei; (2) the large Telescope, provided by the University of Chicago, is designed to measure the energy spectra and the charge/mass composition of solar energetic particles from helium through nickel over the energy range 5–250 MeV/nucleon.

The detectors employ Li-drifted silicon detectors and position-sensitive detectors.

Mission Strategy:

EHIC will be flown as a secondary payload on either ATN–I or ATN–J.

Enabling Technology Development: None

Points of Contact:

Program Manager:	Rick Howard
Program Scientist:	Vernon Jones
NASA Center:	GSFC
Project Manager:	Bob Wales
Project Scientist:	TBD

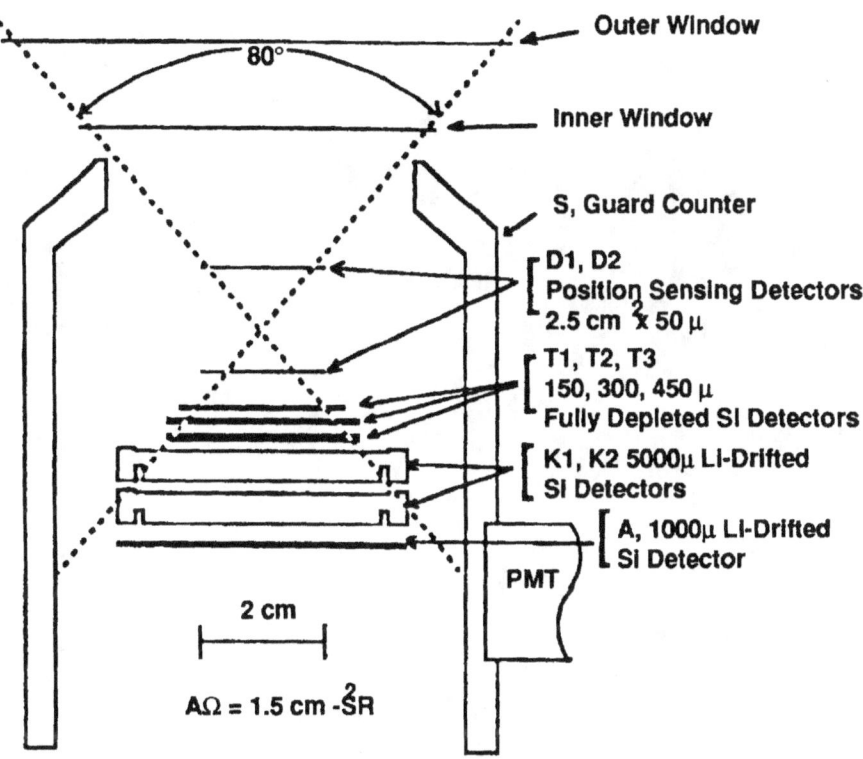

EHIC large telescope

4.12 Atmospheric Laboratory for Applications and Science (Atlas-1)

Target: Sun and Earth atmosphere

Orbit: 300 km x 300 km, 57° inclination

Mission Duration: 7 days

Mission Class: Small (shuttle attached payload)

Mass: 9,000 kg

Launch Vehicle: STS

Theme: Solar physics and Earth atmosphere reactions

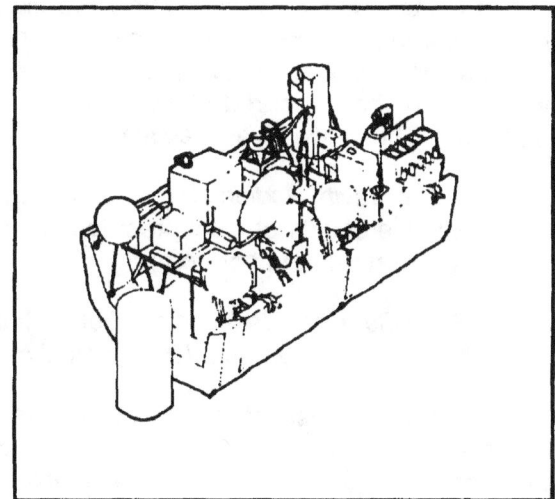

Science Objectives:

To measure long-term changes in the total energy radiated by the Sun, to determine the variability in the solar spectrum, and to measure the global distribution of key molecular species in the middle atmosphere. Such measurements are needed because even small changes in the Sun's total irradiance or its spectral distribution can have a significant impact on the Earth's climate and environment. Additional objectives are to differentiate manmade from natural perturbations in the Earth's atmosphere and to provide absolute calibrations for solar monitoring instruments on free-flying spacecraft.

Spacecraft:

Type: STS attached payload (2 pallet and igloo)

Special Features: Inertially fixed for solar viewing or nadir pointing for Earth viewing

Special Requirements: TBD

Instruments:

The core payload for the Atlas missions includes six proven instruments in a continuing effort to monitor variations in the total solar irradiance and the solar spectrum with state-of-the-art precision and to characterize the response of the Earth's atmosphere to changes in the incident solar energy. A series of flights is planned so that similar measurements can be made over a complete 11-year cycle of solar activity.

Two of these investigations will measure the solar constant, the total solar irradiance from far-ultraviolet through infrared wavelengths at a distance of one astronomical unit, with state-of-the-art accuracy and precision. Both instruments are heat detectors, or pyroheliometers, which control the amount of electrical power supplied to a heater to maintain a constant temperature in a cavity. The solar constant is determined by observing the difference in electrical power as a shutter opens and closes the aperture that allows sunlight into the cavity. These measurements will be used to measure short-term fluctuations in the solar constant during a flight and, by measurements taken during a number of flights, to sustain the long-term absolute accuracy of the solar irradiance data base at the required ± 1-percent level. The two types of pyroheliometers will be compared with each other and with similar experiments on free-flying satellites. These two investigations are:

- *Active Cavity Radiometer.* This instrument has three independent cavities and is capable of 0.01 percent precision in a single 2-minute measurement. (R.C. Wilson, Jet Propulsion Laboratory)

- *Solar Constant Radiometer.* Better than 0.5 percent precision is expected for this two-cavity instrument. (D. Crommelynck, Institut Royal Meteorologique, Belgium)

Another pair of investigations will observe the spectral distribution of the Sun's energy output to determine which wavelength ranges of the solar spectrum are involved in the variability of the solar constant. The ultraviolet wavelengths from 170 to 210 nm are of particular interest because they are absorbed by oxygen and ozone molecules in the stratosphere. The measurement goal for absolute accuracy is to approach the limits set by transfer standards (up to ±5 percent at some wavelengths) and that for relative intensities of line and continuum emission is ±1 percent. These two investigations are:

- *Solar UV Spectral Irradiance Monitor (SUSIM).* The absolute solar flux at wavelengths from 120 to 400 nm is measured by two identical double-dispersion, scanning spectrometers that share a common calibration light source and a set of seven detectors. (G.E. Brueckner, Naval Research Laboratory)

- *Solar Spectrum Irradiance Monitor (SSIM).* Three spectrometers cover the ultraviolet, visible, and near-infrared spectral ranges, that is, wavelengths of 170 to 3,200 nm. Each spectrometer has a double-dispersion monocrometer for wavelength scanning, a detector, and a calibration light source. (G. Thuiller, Service d'Aeronomie du CNRS, France)

Other core investigations will look for long-term changes in the composition of the Earth's atmosphere. They are:

- *Atmospheric Trace Molecule Spectroscopy (ATMOS).* ATMOS is a Fourier transform infrared spectrometer that views the Earth's limb and records solar spectra with absorption features caused by atmospheric constituents. From the spectra, the compositional structure of the upper atmosphere and the related physics and chemistry can be determined over a range of latitudes, longitudes, and altitudes. (C.B. Farmer, Jet Propulsion Laboratory)

- *Imaging Spectrometric Observatory (ISO).* Five spectrometers covering extreme UV to IR wavelengths will obtain complete spectra of atmospheric airglow. (M.R. Torr, NASA, Marshall Space Flight Center)

Section 4 Approved Missions

- *Shuttle Solar Backscatter Ultraviolet Experiment (SSBUV).* The vertical distribution of ozone and some other minor constituents in the Earth's atmosphere can be deduced from the measurements of backscattered solar radiation made by this instrument in narrow ultraviolet wavelength bands. SSBUV will be used to calibrate similar instruments on polar meteorological satellites and to determine long-term trends in ozone concentrations. (Hilsenrath, Goddard Space Flight Center)

- *Microwave Atmospheric Sounder (MAS).* This investigation will measure the microwave absorption lines of water, oxygen, ozone, and other trace gases at high spectral resolution. The data will be used to deduce vertical profiles of gas concentration, kinetic temperature, and pressure for altitudes between 20 and 90 km. (Hartman, MPI-Aeronomy, West Germany)

The following reflight investigations will be accommodated only on Atlas-1:

- *Space Experiments with Particle Accelerators (SEPAC).* (T. Obayashi, University of Tokyo, Japan)

- *Atmospheric Emission Photometric Imaging (AEPI).* (S.B. Mende, Lockheed Palo Alto Research Laboratories)

- *Far Ultraviolet Astronomy, Using the FAUST Telescope.* (C.S. Bowyer, University of California at Berkeley)

- *Grille Spectrometer.* (M. Ackerman, IASB, Belgium)

- *Investigation on Atmospheric H and D Through the Measurement of Lyman-Alpha (ALAE).* (J.L. Bertaux, CNRS, France)

Some additional investigations are being considered for Atlas-2 and any follow-on flights in the series.

Mission Strategy:

Atlas 1 will be launched on the STS in May 1991. The first Atlas mission will use two Spacelab pallets and an igloo to accommodate a core payload of solar and atmospheric monitoring instruments plus reflights of some Spacelab investigations. Later missions at 1- to 2-year intervals will have a single pallet. The Orbiter orientation will be either inertially fixed so that selected instruments are pointed at the Sun or nadir-pointed for observations of the Earth's atmosphere. The orbit must have solar occultations so that absorptions in the solar spectrum caused by trace molecules in the atmosphere can be detected by the ATMOS instrument, an infrared spectrometer with a mirror system to track the Sun. Command, control, and data handling support for the experiments are provided by Spacelab's avionics located in the igloo. The crew will work at the aft flight deck, which has the displays and controls needed to conduct Atlas's investigations.

Enabling Technology Development: None

Section 4 Approved Missions

Points of Contact:

Program Manager:	Lou Demas
Program Scientist:	Dixon Butler
NASA Center:	MSFC
Project Manager:	Tony O'Neil
Project Scientist:	Marsha Torr

Section 4 Approved Missions

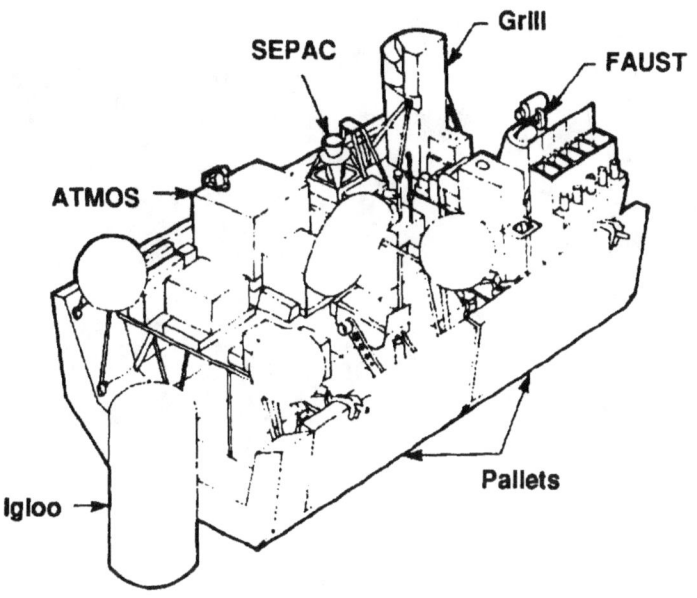

Atmospheric Laboratory for Applications and Science (Atlas-1)

Section 4 Approved Missions

4.13 Fast Auroral Snapshot Explorer (FAST)

Target: Earth auroral zone

Orbit: 350 km x 4200 km, 83° inclination

Mission Duration: 2 years

Mission Class: Small Explorer

Mass: TBD

Launch Vehicle: Scout-class vehicle
September 1993

Theme: Magnetospheric physics

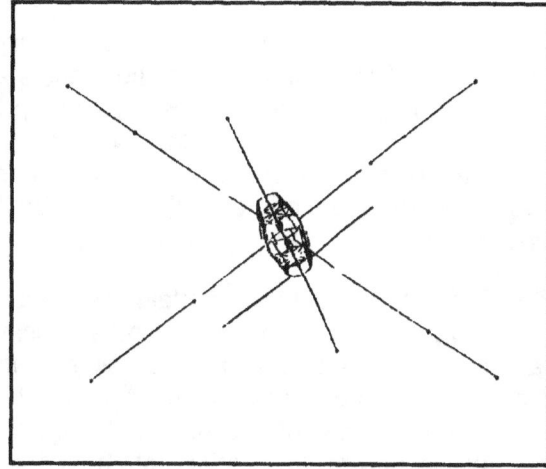

Science Objectives:

An innovative, high time-resolution set of coordinated instruments that will examine the electrodynamic causes of intricately complex auroral displays. Of special importance will be an attempt to reveal how electrical and magnetic forces guide and accelerate electrons, protons, and other ions in the auroral regions.

Spacecraft:

    Type: Small Explorer

    Special Features: TBD

    Special Requirements: TBD

Instruments:

Five instruments based on previous satellite and rocket experience:

Investigation	Expt code	Mass (kg)	Data rate (bps)	Principal Investigator
Fast Electron Spectrograph	FES	25.0		C. Carlson (Berkeley)
Electron Electrostatic Analyzer	EESA	7.5		C. Carlson
Toroidal Ion Mass-Energy-Angle Spectrograph	TIMEAS	5.0		C. Carlson
Ion Electrostatic Analyzer	IESA	1.5		C. Carlson
Electric & Magnetic Field Instrument		1.9		C. Carlson

Mission Strategy:

The FAST mission will be launched on a Scout-class vehicle in June 1993. This investigation of auroral processes expands upon a wide range of plasma phenomena discovered on previous satellite and rocket missions. Its high-resolution measurements will greatly extend the observational capability of sounding rockets and should make a significant contribution to understanding the basic physics of auroral particle acceleration. The FAST spacecraft will operate in the natural plasma laboratory above the Earth's auroral zones. This domain of the upper atmosphere is where the neutral atmosphere contacts the plasma-dominated solar system environment of the Earth. Energy and matter flow through this region, exciting the upper atmosphere into luminous displays controlled by electrical and magnetic forces.

The FAST instruments will contain new sensors capable of detecting the flows of various types of matter, electrons, protons, and other ions, with greater sensitivity, discrimination, and much faster sampling than previously possible. Other sensors will measure the electrical and magnetic forces and simultaneously correlate these forces with their effects on the electrons and ions at altitudes of 300 km to 3500 km, the source region for much of the energy that appears as auroral light emitted at about 100 km. These observations will be complimented by data from other spacecraft at higher altitudes, which will be observing fields and particles and photographing the aurora from above, thus placing FAST observations in global context. At the same time, auroral observatories and geomagnetic stations on the ground will provide measurements on how energetic processes that FAST observes affect the Earth.

Enabling Technology Development: None

Points of Contact:

Program Manager:	Dave Gillman
Program Scientist:	Tom Armstrong
NASA Center:	GSFC
Project Manager:	Ron Askins
Project Scientist:	Dan Baker

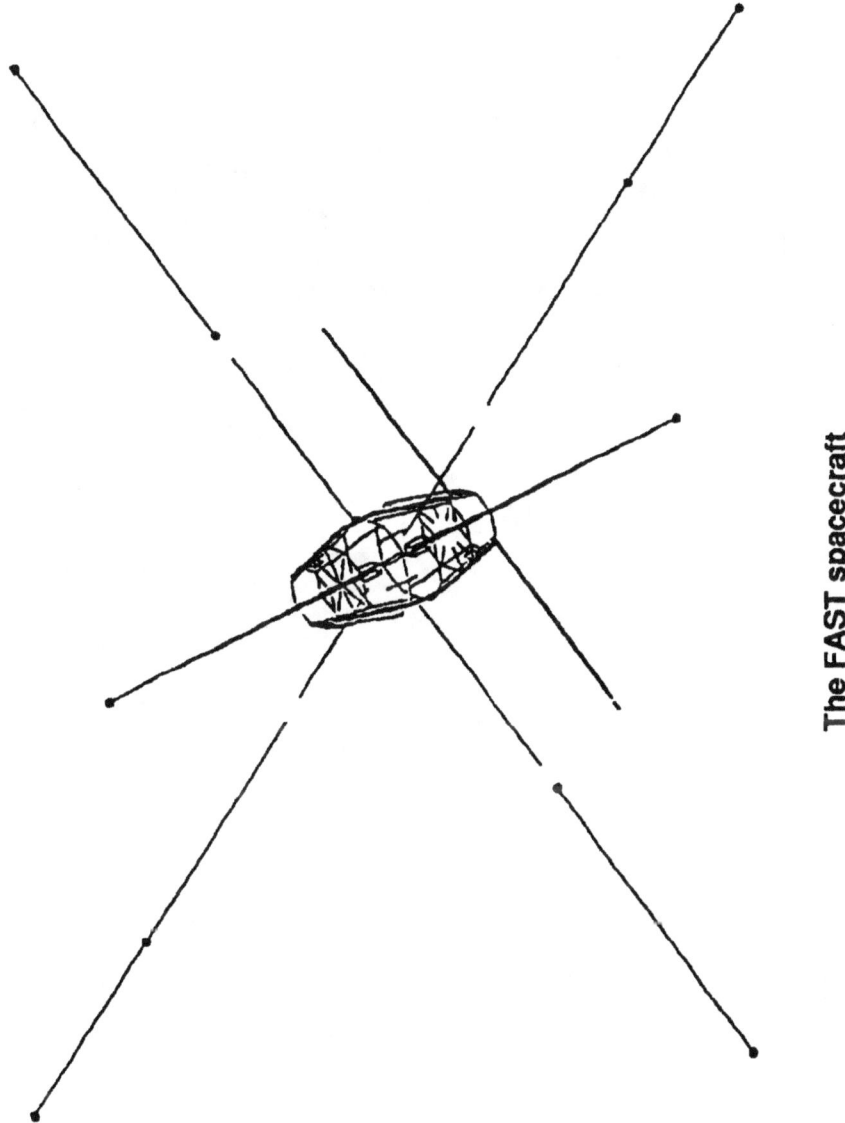

The FAST spacecraft

4.14 Advanced Composition Explorer (ACE)

Target: Heliosphere

Orbit: L1 halo

Mission Duration: 5 years

Mission Class: Explorer

Mass: 633 kg (total)
 105 kg (payload)

Launch Vehicle: Delta II

Theme: Observation of solar, interplanetary, interstellar, and galactic particles

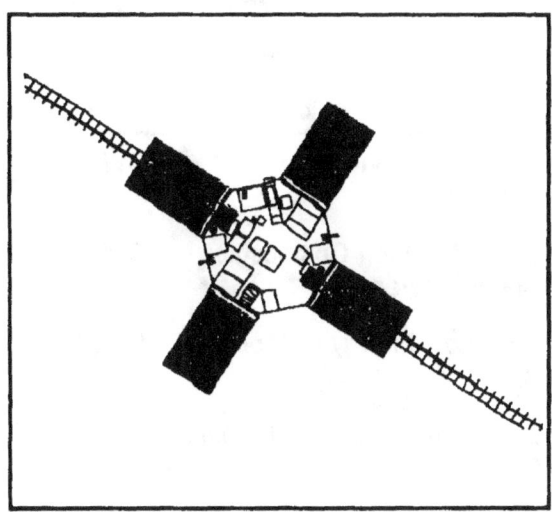

Science Objectives:

The prime objective of ACE would be to determine and compare the elemental and isotopic composition of several distinct samples of matter, including the solar corona, the interplanetary medium, the local interstellar medium, and galactic matter. This objective is approached by performing comprehensive and coordinated determinations of the elemental and isotopic composition of energetic nuclei accelerated on the Sun, in interplanetary space, and from galactic sources. These observations would span five decades in energy, from solar wind (several hundred MeV per nucleon) to galactic cosmic rays (1 keV per nucleon) and would cover the element range from 1H to 40Zr. The comparison of these samples of matter would be used to study the origin and subsequent evolution of both solar-system and galactic material by isolating the effects of fundamental processes that include nucleosynthesis, charged- and neutral-particle separation, bulk plasma acceleration, and the acceleration of suprathermal and high-energy particles.

Spacecraft:

 Type: Explorer, spin stabilized along Sun-Earth axis
 Special Features: TBD
 Special Requirements: TBD

Instruments:

The scientific instrumentation for the ACE spacecraft consists of a coordinated set of 6 sensor systems with unprecedented resolution and sensitivity: a Solar Wind Ion Mass Spectrometer (SWIMS), a Solar Wind Ion Composition Spectrometer (SWICS), an Ultra-Low Energy Isotope Spectrometer (ULEIS), a Solar Energetic Particle Ionic Charge Analyzer (SEPTICA), a Solar Isotope Spectrometer (SIS), and a Cosmic Ray Isotope Spectrometer (CRIS). In addition, there

Section 4 Approved Missions

are three small subsystems: Solar Wind Electron, Proton, and Alpha Monitoring (SWEPAM), energetic Electron, Proton, and Alpha Monitoring (EPAM), and a Magnetometer (MAG).

Typical Energy Intervals (in MeV/nuc) for Multiparameter Analysis

Investigation	Expt code	Mass (kg)	Data rate (bps)	Principal Investigator
Solar Wind Ion Mass Spectrometer	SWIMS	6.8	505	E. Stone (Cal Tech)
Solar Wind Ion Composition Spectrometer	SWICS	5.5	500	E. Stone
Ultra-Low Energy Isotope Spectrometer	ULEIS	17.0	1000	E. Stone
Solar Energetic Particle Ionic Charge Analyzer	SEPICA	16.0	600	E. Stone
Solar Isotope Spectrometer	SIS	19.5	2000	E. Stone
Cosmic Ray Isotope Spectrometer	CRIS	20.6	462	E. Stone
Solar Wind Electron, and Alpha Monitoring	SWEPAM	6.7	1000	E. Stone
Energetic Electron, Proton, and Alpha Monitoring	EPAM	5.0	160	E. Stone
Magnetometer	MAG	4.7	300	E. Stone

Atomic Number	Isotopes	Elements	Integral fluxes of elements	Ionic charge states
Z = 1		0.0001–50		
Z = 2	0.0001–120	0.0001–180	>180	0.0001–2.5
3<Z<8	0.0001–220	0.0001–320	>320	0.0001–2.5
9<Z<14	0.0001–340	0.0001–520	>520	0.0001–2.3
Z>40	0.0001–1			

(Exploratory; for Z> 40, only even-Z elements may be possible.)

Parameter	SWIMS	SWICS	ULEIS
Geometry factors ($cm^2 sr$):	0.013	0.009	0.95
Mass resolutions	Ne: 0.10	0.10+	0.15
(rms 0 in amu)	Fe: 0.20	0.15+	0.40

For SWIMS and SWICS the active area in cm^2 is shown. For SWICS and SEPICA the rms charge res.

Mission Strategy:

The ACE study payload includes six high-resolution spectrometers, each designed to provide optimum charge, mass, or charge-state resolution in its particular energy range, and each having a geometry factor optimized for the expected flux levels, so as to provide a collecting power a factor of 10 to 1000 times greater than previous or planned experiments. The flux dynamic range of these instruments will be sufficient to perform measurements under all solar wind flow conditions and during both large and small solar-particle events, including 3 He-rich flares. Magnetic field, solar wind electrons, and solar flare electrons will also be measured.

The ACE spacecraft is based on the Charge Composition Explorer (CCE) built by JHU/APL for the AMPTE program. It is a spinning spacecraft with its spin axis aligned to the Earth/Sun axis. The ACE launch weight will be ~633 kg (including 105 kg scientific instruments and 184 kg propellant). It will be launched into an L1 libration point (240 Re) orbit with a Delta-class expendable launch vehicle. Telemetry will be 6.7 kbps average (tape recorder storage with daily readout to DSN).

Enabling Technology Development: None

Points of Contact:

Program Manager:	TBD
Program Scientist:	Vernon Jones
NASA Center:	GSFC
Project Manager:	TBD
Project Scientist:	TBD

Section 4 Approved Missions

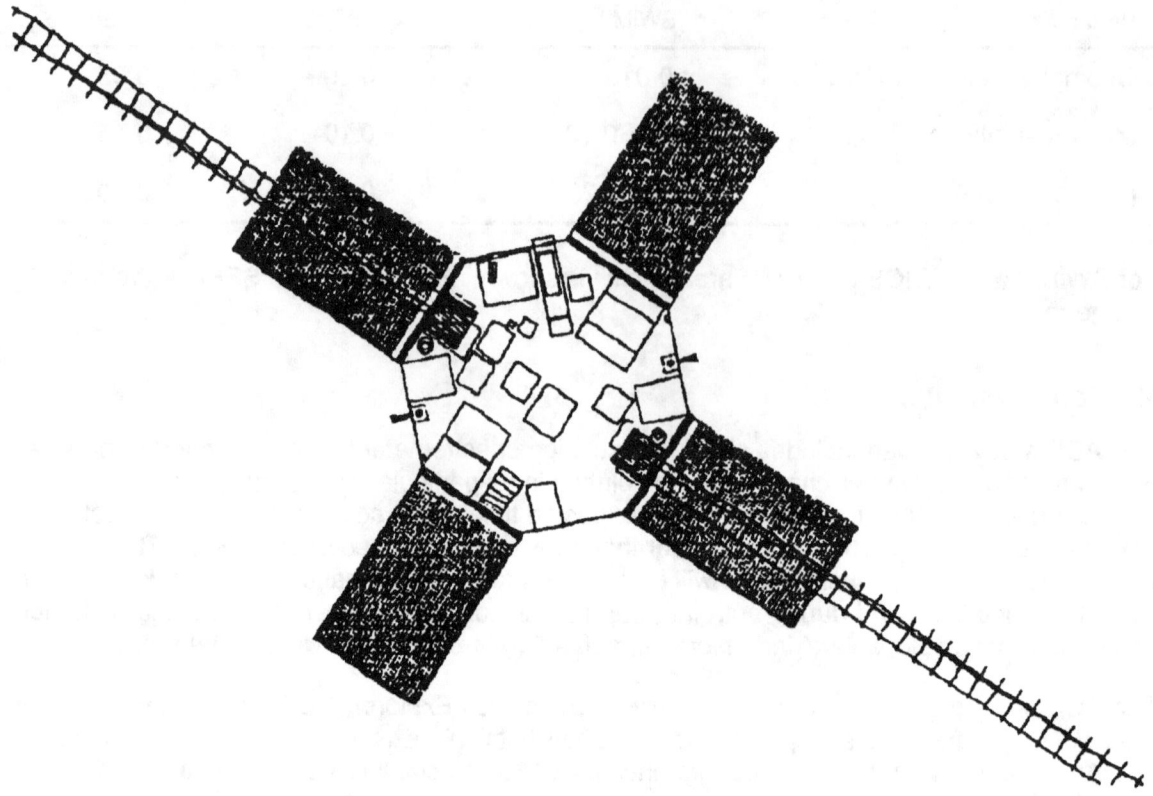

Advanced Composition Explorer (ACE)

Section 5
Planned Missions

5.1 Neutral Environment With Plasma Interaction Monitoring System (NEWPIMS)

Target: Space Station environment

Orbit: Space Station attached payload

Mission Duration: Continuous

Mission Class: Small

Mass: TBD

Launch Vehicle: STS (TBR)

Theme: Space Station environment monitoring

Science Objectives:

NEWPIMS will monitor the charged and neutral particle environment.

Spacecraft:
- **Type:** Space Station attached payload
- **Special Features:** TBD
- **Special Requirements:** TBD

Instruments: TBD

Mission Strategy:

The NEWPIMS prototype would be launched early in the assembly sequence and will be attached to the Space Station as an ongoing experiment. Five more units would be added during the space station assembly phase to monitor the space station's environment at different attached locations

Enabling Technology Development: None

Points of Contact:
- **Program Manager:** Bill Roberts
- **Program Scientist:** Vernon Jones
- **NASA Center:** MSFC
- **Project Manager:** TBD
- **Project Scientist:** TBD

Section 5 Planned Missions

5.2 Heavy Nuclei Collector (HNC)

Target: Cosmic rays

Orbit: Space Station orbit

Mission Duration: 2 1/2 years

Mission Class: Space Station attached payload

Mass: 3400 kg

Launch Vehicle: STS (TBR)

Theme: Cosmic ray physics

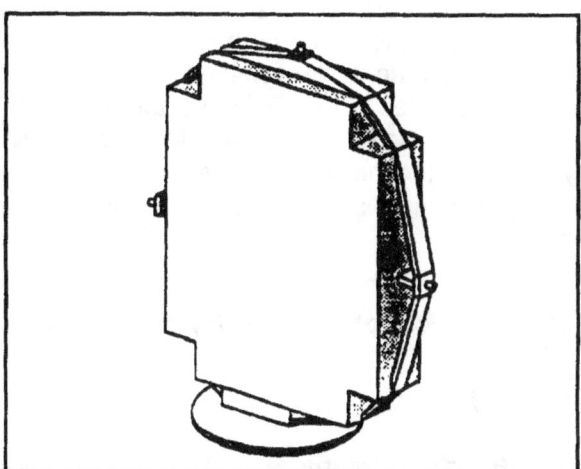

Science Objectives:

The HNC is an array of passive glass detectors to study the origin of ultra-heavy cosmic rays. It will determine the relative abundances of odd- and even-charged cosmic ray nuclei, search for highly ionizing particles, such as super-heavy elements, and attempt to determine the sources of ultra-heavy cosmic rays.

Spacecraft:

Type: Space Station attached payload

Special Features: TBD

Special Requirements: TBD

Instruments:

Array of passive glass detectors.

Mission Strategy:

HNC has been conditionally selected, flying only if UHRXS does not. The HNC is completely passive and requires no electrical power. It consists of trays, stacked with phosphate glass. The glass sheets (14 sheets per tray) record the tracks of the cosmic rays as they either pass through or are captured in the glass and disintegrate. The glass detectors are recovered and returned to Earth for analysis. The analysis consists of etching in the individual glass sheets and measuring the latent cosmic ray tracks, to derive the mass and energy data. This technique, which originally used plastic detectors, was developed by the HNC principal investigator. Phosphate glass has better nuclear charge resolution than plastic detectors and is less sensitive to orbital thermal variations experienced by Space Station Freedom. Glass is also less sensitive to overexposure to space radiation, and therefore can remain in space longer.

Section 5 Planned Missions

The glass stacks are mounted like windows in the trays so that both sides of the sheets are exposed to space. The HNC will be mounted in the anti-Earth direction. The HNC data will be used by the investigators to better understand the synthesis of cosmic rays whose nuclear masses range from hydrogen to, and possibly exceeding, uranium. The heaviest particles (actinides) are short-lived on a cosmic scale, so their relative abundances can yield major new information on the nucleosynthesis mechanisms and the astronomical environment in which they were created. The HNC detector array will have cumulative collection power equivalent to flying 32 m^2 of detectors in space for 4 years. With only 14 layers of glass as compared to 120 layers of plastic, the post-flight mission can be accomplished with only two institutions doing the etching, scanning, and measuring, and with one other institution participating in the interpretation.

Enabling Technology Development: None

Points of Contact:

Program Manager:	Lou Demas
Program Scientist:	Vernon Jones
NASA Center:	MSFC
Project Manager:	Rein Ise
Project Scientist:	TBD

5.3 Orbiting Solar Laboratory (OSL)

Target: Solar atmosphere

Orbit: Sun-synchronous, 97.4°, 510 km

Mission Duration: 3 years

Mission Class: Moderate

Mass: 3400 kg

Launch Vehicle: Delta II (1997)

Theme: Solar physics

Science Objectives:

The objective of OSL is to provide the angular resolution, sensitivity, and stability to permit study of the fundamental interactions of solar surface wave and flow fields with the magnetic field at the scale of 100 km as measured at the Sun's surface. OSL will specifically investigate:

- Photospheric magnetic and velocity fields (Visible)
- Chromospheric and transition zone spectroscopy (UV)
- Coronal imaging and spectroscopy (XUV, X-ray)

OSL science objectives will complement results from other upcoming solar missions such as the NASA joint endeavor with the European Space Agency (ESA)—the Solar and Heliospheric Observatory (SOHO). SOHO will study solar corona and solar wind phenomena and the interior dynamics of the Sun from the core up to the photosphere. Synergistic SOHO and OSL observations of solar oscillations, for example, will enhance the value of the data collected. Other missions that will benefit from OSL operation include major international programs such as Global Geospace Science (GGS) and Cluster.

Spacecraft:

    Type: 3-axis stabilized
    Special Features: TBD
    Special Requirements: TBD

Instruments:

OSL Main Telescope

- 1 m diameter Gregorian reflecting telescope
- 24 m effective focal length
- commandable secondary mirror
- 0.13 arc-second spatial resolution
- 3.9 arc min unvignetted field of view

Photometric Filtergraph (H. Zirin, California Institute of Technology)

- 160 x 160 arc-second field filter camera
- 2300–6687 Å, low spectral resolution

Tunable Filter/Magnetograph (A. Title, Lockheed Palo Alto Research Laboratory—LPARL)

- 160 x 160 arc-second field
- 4600–6563 Å medium spectral resolution

Kiepenheuer Institute Solar Spectrograph (E. Schröter, Kiepenheuer Institute, Federal Republic of Germany)

- 0.1 x 80 arc-second field
- 2800–8540 Å, high spectral resolution

High-Resolution Telescope and Spectrograph (G. Brueckner, Naval Research Laboratory)

- 30 cm diameter Gregorian reflecting telescope
- Spectrograph (1200–1700 Å)
- UV slit image display (1550 Å)
- Visible slit image display (6553 Å)

X-Ray Ultraviolet Imager (E. Antonucci, Italian Space Agency, USAF Geophysics Laboratory)

- 45–335 Å, high-resolution imaging
- Full-Sun imaging capability
- Multilayer coatings to isolate different wavelengths

The largest telescope on OSL will be a 1-meter diameter optical telescope, with a spatial resolution of about 100 km on the solar surface, which will supply visible light to all three instruments in the CIP. A Photometric Filtergraph from the California Institute of Technology, operating between the near-infrared and the near-ultraviolet wavelengths, will be particularly useful in studying the heating of the upper photosphere. A second CIP instrument, the Tunable Filtergraph from the Lockheed Palo Alto Research Laboratory, will map motions and magnetic fields in the photosphere to help study the interaction between velocities and magnetic fields—a key goal of OSL. The third instrument—the Kiepenheuer Institute Solar Spectrograph from Germany—will provide measurements of magnetic field, velocity, and turbulence in the photosphere and chromosphere.

The High Resolution Telescope and Spectrograph (HRTS), provided by the U.S. Naval Research Laboratory, produces spatial resolution of 175 km through a 30-cm aperture telescope. It has three focal plane instruments. One spectrograph, operating at ultraviolet wavelengths, will reveal information on atmospheric heating and motions in the high chromosphere and the transition zone to the corona. A second spectrograph will obtain data to study thermal structures at the base of the chromosphere. The third HRTS instrument is a filter imager operating at the red H-alpha wavelengths, provided to ensure precise co-registration of data with the CIP data from the main optical telescope.

The final science instrument on OSL is the X-Ray Ultraviolet Imager (XUVI), a joint effort between the Italian Space Agency and the U.S. Air Force. The current concept comprises a high-resolution imaging (HRI) capability, with two identical Newtonian telescopes. Each has a primary mirror with new multilayer coatings that will provide, for the first time, the ability to isolate the higher temperature extreme ultraviolet (EUV) and soft X-ray emissions from the solar corona, with a spatial resolution better than 400 km. A full Sun imager (FDI) is being considered to define observational targets, as well as for studying large-scale phenomena.

Mission Strategy:

OSL will be a three-axis-stabilized spacecraft, with a meter-class telescope and two smaller co-observing instruments. It will be launched into a Sun-synchronous orbit that will provide nine months of continuous solar observations per year. OSL is complementary to the Solar and Heliospheric Observatory (SOHO), Solar-A, and the NOAA X-ray imaging monitor in the objective of understanding the solar cycle.

With sufficiently high spatial and temporal resolution across a wide spectral range OSL also has the requisite instrumentation and mission architecture to discover the precursor solar phenomena that reliably indicate that a solar flare is about to occur. Thus, if OSL is launched by the beginning of the rise phase of the next solar activity cycle in 1997, sufficient knowledge of flare precursor phenomena should be available to permit the design of special instruments just for that purpose. Such instrumentation could support long-term manned presence on the Moon and the exploration of Mars.

Unlike the Skylab mission, which collected data on film for later analysis, OSL will use high-resolution Charge-Coupled Device (CCD) cameras. The images are stored electronically for transmission to the ground and analysis by OSL investigators. The OSL mission will use NASA's Tracking and Data Relay Satellite System (TDRSS) for satellite communications. Spacecraft handling will be controlled by the Payload Operations Control Center (POCC) at the NASA Goddard Space Flight Center (GSFC).

A Science Data and Operations Center (SDOC) to be located at GSFC will plan and schedule observations and OSL pointings. During each observation period, commands sent to OSL will control instrument settings and the final positioning of the target area in the field of view. Data collection capability permits science operations for about nine hours each day, for 250 days of the year. The data collected will be received by the SDOC, via the POCC, for processing, analysis, and storage. The SDOC will also offer investigators "quick-look" data products to support time-critical observation of events and analysis of newly acquired data within 48 hours

Section 5 Planned Missions

of receipt by SDOC. The SDOC will also interface with geographically remote workstations for data analysis and investigator interaction with observations, using national computer networks.

OSL will be operated as a solar physics facility that will support not only Principal Investigators (PI's), Co-Investigators (Co-I's), and Facility Scientists, but will also support programs of competitively selected theorists, Guest Observers (GO's), and Guest Investigators (GI's). The observation program will be coordinated with ground-based solar observatories around the world and with other concurrent solar missions in space, such as the Solar and Heliospheric Observatory (SOHO) to be launched early in 1995.

Enabling Technology Development:
- CCD's with appropriate spectral response

Points of Contact:

Program Manager:	Rick Howard
Program Scientist:	Bill Wagner
NASA Center:	GSFC
Project Manager:	Roger Mattson
Project Scientist:	Dan Spicer

5.4 Ultra-High Resolution Extreme Ultraviolet Spectroheliograph (UHRXS)

Target: Sun

Orbit: Space Station

Mission Duration: 1 year

Mission Class: Space Station attached payload

Mass: 270 kg

Launch Vehicle: STS (TBR)

Theme: High-resolution spectroscopy of EUV lines

Science Objectives:

Observe and study the solar chromosphere, corona, and corona/solar wind interface in the extreme ultraviolet and at a resolution sufficient to discern the physical structures controlling the dynamics of these solar elements.

UHRXS is an ultra-high resolution XUV spectroheliograph which will address fundamental scientific problems relating to several solar phenomena which include:

- The morphology of the fine structure of the solar chromosphere/corona interface, including the "chromospheric network," spicules, prominences, cool loops, and the magnetic field

- The structure, energetics, and evolution of high temperature coronal loops

- The large scale structure and dynamics of the corona, including the solar wind interface (*e.g.*, polar plumes) the coronal magnetic field, and coronal mass ejections

- Solar flares, especially the evolution of the pre-flare state, the nature of the impulsive energy release, and the evolution of the post-flare loops.

Spacecraft:

Type: Space Station attached payload

Special Features: TBD

Special Requirements: Space Station pointing mechanism

Section 5 Planned Missions

Instruments: Ritchey-Cretien telescopes (8)

Mission Strategy:

UHRXS consists of eight identical Ritchey-Cretien telescopes covering the 40Å to 1600Å spectral range to be mounted on the Space Station Freedom Payload Pointing System (PPS). Image-motion compensation is required for solar pointing to allow sub-arc-second resolution. The XUV images are recorded on high-resolution 70mm format film. Film canisters, which are approximately 14" in diameter, are robotically exchangeable. The telescopes employ multi-layer optics technology and Multi-Anode Microchannelplate Array (MAMA) detectors.

Enabling Technology Development:

- Normal-incidence mirror coatings
- High-precision pointer

Points of Contact:

Program Manager:	Lou Demas
Program Scientist:	Dave Bohlin
NASA Center:	MSFC
Project Manager:	Rein Ise
Project Scientist:	TBD

5.5 Astromag

Target: Galactic cosmic rays

Orbit: Space Station Freedom

Mission Duration: 4 years

Mission Class: Space Station attached payload

Mass: TBD

Launch Vehicle: STS

Theme: Particle astrophysics/space physics

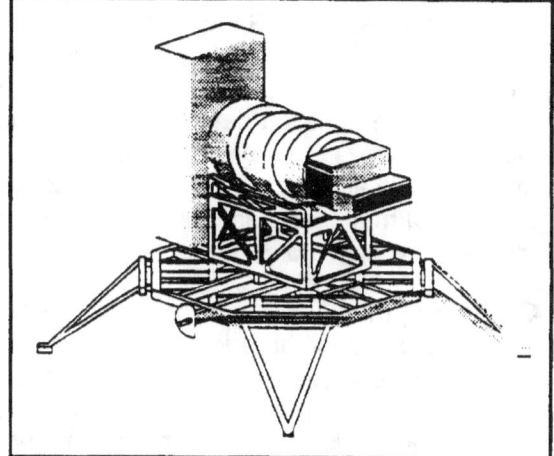

Science Objectives:

The objectives of the selected Astromag experiments are to:

- Investigate the origin and evolution of matter in the Galaxy.
- Search for antimatter and investigate the nature of dark material.
- Determine the origin of extremely energetic particles.

Spacecraft:

Type: Space Station attached payload
Special Features: TBD
Special Requirements: TBD

Instruments:

These instruments will be flown on the Astromag facility. They are Wizard, measuring cosmic rays; SCIN/MAGIC, measuring the spectra, composition and interaction of nuclei above 10 TeV; and LISA, a large isotope spectrometer for Astromag.

	Lifetime	Mass	Size	View direction
Wizard	2 1/2 yrs	2593 kg	4.5 x 1.5 x 3m	Celestial

Wizard will investigate cosmic ray antiprotons, positrons, and light nuclei (helium, lithium) and search for primordial antimatter. These studies are among the highest priority goals of particle astrophysics research, which was cited as key to understanding violent events in the universe by the National Research Council's Astronomy Survey Committee. Of crucial importance are

Section 5 Planned Missions

the relative abundances and energy measurements of antiprotons and positrons, which are predominantly secondary in nature.

The Wizard experiment utilizes an array of particle detectors mounted in the strong magnetic field at one end of the Space Station Freedom Astromag facility. Wizard is modular and consists of a pair of time-of-flight detectors, a tracking system, transition radiation detectors, and a calorimeter. The time-of-flight detectors and the tracking system identify primordial matter. The transition radiation detector and the calorimeter provide the means for separately identifying light particles (*e.g.*, electrons and positrons) from heavier particles (protons and antiprotons). The total instrument constitutes an array of detectors capable of identifying low-mass particles of matter and antimatter over a wide range of energies. Robotic installation and removal are planned.

Quantifying the energy spectra will answer many questions in cosmic ray physics and particle astrophysics about the origin of antiprotons and positrons, selection and acceleration of particles, propagation history of individual components, effect of discrete sources, re-acceleration by interstellar shock waves, and gas field systems in the Galaxy. The Wizard data will provide information on the antimatter/matter asymmetry of the universe, the mini black holes, and the role of the Grand Unified Theories in the early universe, all of which are relevant to both cosmology and elementary particle physics.

	Lifetime	Mass	Size	View direction
SCIN/MAGIC	3 mo each (for 2 pallets)	(1200 kg) x 2	(1.3x1.1x0.75m) x 2	Celestial

SCIN is a high-energy nuclear physics/particle astrophysics experiment which will be mounted on the Space Station Freedom Astromag facility. The goal for SCIN is to measure the composition and spectra of the cosmic ray nuclei above 10^{14} eV and, using the magnetic field, to study the characteristics of nucleus-nucleus interactions above 10^{12} eV per nucleon.

The SCIN apparatus consists of two pallets, each with two passive emulsion chamber detectors that contain nuclear track emulsions, CR 39 plastic track-etch detectors, X-ray films, and lead, tungsten, or other inert materials. One chamber is similar to those flown on balloon-borne experiments to study cosmic ray composition and interactions. The other, which contains a low-density target section designed to measure charge signs and transverse moments in the high fields of Astromag, is similar to a chamber exposed at the CERN heavy ion accelerator. Thermal control is achieved with coatings, insulation blankets, and electrical heaters. The instrument looks in the anti-Earth direction. Two 90-day exposures of the two pallet configurations are planned. SCIN is designed for robotics installation and removal of the pallets. The emulsion chamber analysis will follow post-flight measurements of the track-sensitive detectors.

Cosmic ray particles appear to contain nuclei of all elements found on Earth and are known to possess energies up to at least 10^{20} eV per particle. Pervading the Galaxy, and probably beyond, cosmic rays are a probe of high-energy processes occurring in the Galaxy, and they tie together much of the field of high-energy astrophysics. Their origin and acceleration is intimately linked to the origin of the elements themselves; their propagation and confinement depends on the structure of the Galaxy and its magnetic field. The nuclear interactions of the highest-energy particles permit investigations of the particle production process beyond accelerator energies, including a search for the postulated new phase of matter, quark-gluon plasma.

	Lifetime	Mass	Size	View direction
LISA	2 years	2400 kg	4.5 (diameter) x 2.25m	Celestial

LISA is a cosmic ray isotope spectrometer designed to identify isotopes in cosmic rays using the magnet/Cherenkov/time-of-flight method. LISA will conduct high-precision measurements to further our understanding of galactic material from beyond the solar system. The relative abundances of the elements and isotopes in galactic cosmic rays represent a record of the history and samples of matter from other regions of the galaxy, including its synthesis in stars, its acceleration to high energy, and its subsequent nuclear and electromagnetic interactions with the interstellar medium.

LISA consists of a modular combination of scintillating-optical-fiber trajectory detectors, Cherenkov velocity detectors, and time-of-flight scintillators in the magnetic field at one end of the Astromag facility. The scintillators measure time-of-flight and particle charge, Cherenkov counters measure particle velocity, Aerogel Cherenkov counters measure momentum per nucleon, and the scintillating optical fiber trajectory detectors measure two-dimensional track coordinates in the magnetic field. LISA's cylindrical configuration makes optimal use of the high field region near the Astromag magnet coil. By combining the individual measurements, it is possible to identify: 1) nuclear charge; 2) isotopic mass; and 3) whether the particles are matter or antimatter. The preferred observation direction is local zenith, with a fan-shaped beam field-of-view of 110 x 15 degrees.

One of the fundamental yet unanswered questions of cosmology is the degree to which the universe contains antimatter. LISA will search for heavy antinuclei in the cosmic radiation with a sensitivity that is two orders of magnitude better than existing limits. Specifically, LISA will address investigations of the origin and evolution of galactic matter; the acceleration, transport, and time scales of cosmic rays in the Galaxy; and the presence or lack of antimatter in the universe. The Astromag facility offers the opportunity for LISA to extend high-resolution spectroscopic studies of individual cosmic ray elements and isotopes by about an order of magnitude in energy/nucleon. LISA will be "tuned" to resolve isotopes from Be to Zn.

Mission Strategy:

A large superconducting magnet core facility developed jointly with the Italian Space Agency (ASI) and augmented with a variety of specialized detectors, will be flown as an attached payload on the Space Station Freedom. It is expected that second generation experiments will be installed on Astromag after the initially selected experiments are completed. Astromag will take advantage of the space cryogenics technology developed for missions such as IRAS, Spacelab-2, and COBE.

Section 5 Planned Missions

       1998: MAGIC and Wizard up
       1999: LISA up and MAGIC down
       2001: Wizard and LISA down

Enabling Technology Development: None

Points of Contact:

Program Manager:	Rick Howard
Program Scientist:	Vernon Jones
NASA Center:	GSFC
Project Manager:	George Anikis
Project Scientist:	Jonathan Ormes

Section 5 Planned Missions

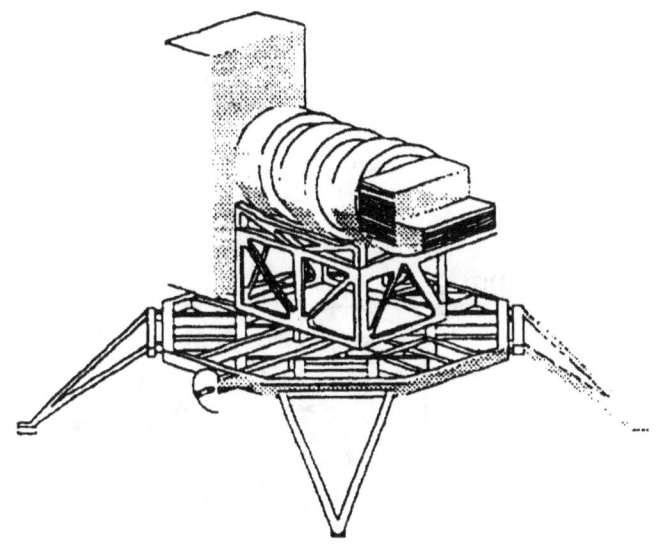

Instruments on Astromag: SCIN/MAGIC

Section 5 Planned Missions

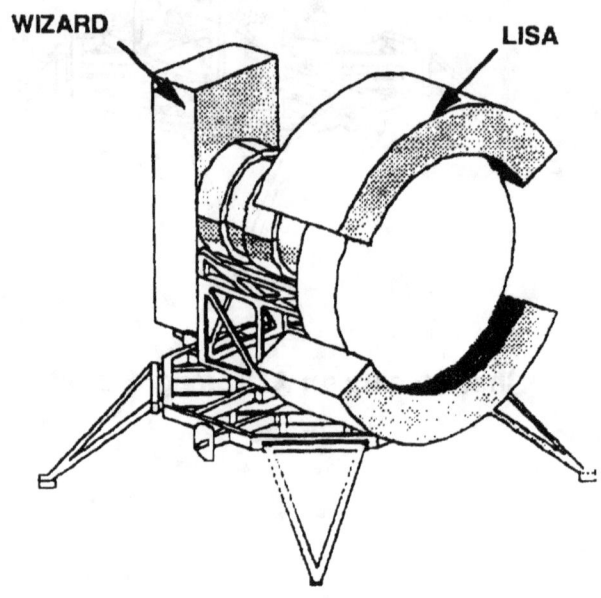

Instruments on Astromag: Wizard and LISA

Section 6
Candidate Future Missions

6.1 Lunar Calorimeter

Target: Lunar-based measurements of high-energy cosmic rays

Orbit: Lunar base

Mission Duration: ≥3 years

Mission Class: Space Exploration Initiative (SEI)

Mass: ~5 tons launch mass

Launch Vehicle: TBD

Theme: Cosmic ray physics

Science Objectives:

Measurement of spectrum and elemental composition of protons through iron up past the "knee" in the cosmic ray spectrum, including cosmic rays with energies from ~10^{14} to 10^{17} eV.

Spacecraft:

    Type: Lunar base facility

    Special Features: Lunar regolith used for bulk of calorimeter material

    Special Requirements: Construction on the Moon using structure transported from Earth and lunar regolith

Instruments:

	Mass	Power	Data rate	Data storage	FOV
Calorimeter	~5 tons lifted from Earth	1 kW	1 kbps	TBD	wide angle

Mission Strategy:

Detector must be assembled on the Moon using lunar regolith for the bulk of the calorimeter material (> 100 tons) together with active detector and structural elements brought from Earth (~5 tons). Ten layers of plastic scintillation counters, each separated by approximately 30–35 cm of lunar regolith, will be viewed by photomultiplier tubes.

Technology Requirements:

- Lunar base

- Method of reducing calorimeter particle "backsplash"

- Detector assembly on the Moon—knowledge of physical properties of lunar regolith

- Calorimeter modules with uniform light collection

Points of Contact:

Science Panel:	Dr. Simon Swordy (312) 702-7835 Laboratory for Astrophysics and Space Research Enrico Fermi Institute University of Chicago 933 East 56th Street Chicago, IL 60637
Discipline:	Cosmic and heliospheric physics
Program Manager:	TBD
Program Scientist:	Dr. Vernon Jones (202) 453-1514
NASA Center:	MSFC
Project Manager:	Carmine DeSanctis (205) 544-0618
Project Scientist:	TBD

Section 6 Candidate Future Missions

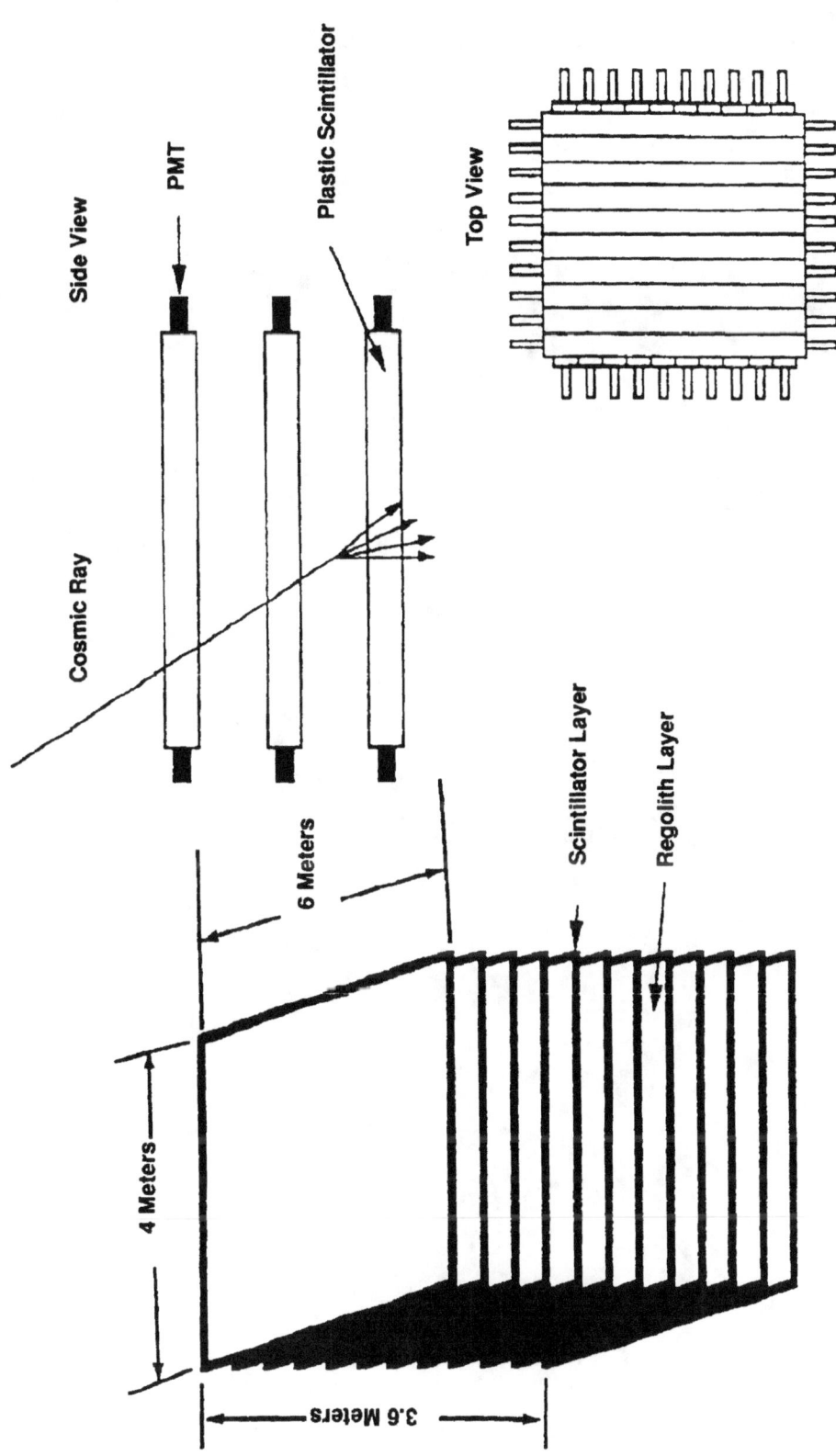

6.2 Neutrino Astrophysics

Target: Neutrinos

Orbit: Lunar base

Mission Duration: Indefinite

Mission Class: Space Exploration Initiative (SEI)

Mass: 500 kg

Launch Vehicle: TBD

Theme: Neutrino astrophysics (cosmic & heliospheric physics)

Science Objectives:

To study neutrino origins without the background "noise" from interactions in the Earth's atmosphere. In particular:

- Search for a diffuse flux of heavy neutrinos.

- Search for a directional flux of weakly interacting massive particle (WIMP) annihilation products from the Earth or the Sun.

Spacecraft:

    Type: Lunar base facility

    Special Features: Location

    Special Requirements: Lunar cavern ~10 m below the lunar surface

Instruments:

	Mass	Power	Data rate	Data storage	FOV
Neutrino telescope (100 m² detector)	500 kg	1 kW	TBD	TBD	Hemisphere

Mission Strategy:

Construct a neutrino telescope in a large natural cavern at least 10 meters below the lunar surface. The roof will provide shielding, the floor will provide target mass for upward-coming neutrinos, and light-weight gas-filled detectors can be deployed on plastic mesh supports.

Technology Requirements:

- Feasibility and safety of natural lunar caverns
- Fast time-of-flight capability and background rejection
- Up-down discrimination and identification of upward-going neutrino events

Points of Contact:

Science Panel:	Dr. Michael L. Cherry (504) 388-8591 Department of Physics Louisiana State University Baton Rouge, LA 70803
Discipline:	Cosmic and heliospheric physics
Program Manager:	TBD
Program Scientist:	Dr. Vernon Jones (202) 453-1514
NASA Center:	MSFC
Project Manager:	Carmine DeSanctis (205) 544-0618
Project Scientist:	TBD

6.3 Tethered Multiprobe ("Pearls on a String")

Target: Earth's lower thermosphere and ionosphere

Orbit: Low Earth orbit, tether mission, inclination: 57.5° & 80°

Mission Duration: TBD

Mission Class: Moderate

Mass: TBD

Launch Vehicle: STS

Theme: Lower thermospheric and ionospheric physics

Science Objectives:

- To study the vertical and horizontal profiles of thermospheric and ionospheric parameters to characterize small (~kms) and medium scale (~ 100s of kms) variability in winds, composition and temperatures.

- To measure vorticity and divergence in the atmosphere on a global scale.

- To map the dynamic structure of the lower E-region on a global scale with 5–10 km resolution.

- To investigate the Sq current system.

- To investigate 3-dimensional current systems in the ionosphere.

- To investigate the transition region between collisionless and collision-dominated regimes of gas kinetics.

- To determine the metallic ion inventory in the lower ionosphere and the interchange roles of winds and E-fields in the formation of intermediate layers.

Spacecraft:

Type: 10 duplicate instruments, ~10 km apart
Special Features: 100–130 km tether
Special Requirements: TBD

Instruments:

	Mass	Power	Data rate	Data storage	FOV
Neutral wind	3 kg	3 W	TBD	TBD	TBD
Ion drift	7 kg	7 W	TBD	TBD	TBD
Neutral, ion, & electron temperatures	3 kg	8 W	TBD	TBD	TBD
Magnetometer	5 kg	4 W	TBD	TBD	TBD

Mission Strategy:

This mission concept would utilize a string of generalized detectors spaced by ~10 km along a non-conducting tether. It has become technologically feasible to deploy tethers up to 130 km in length. Thus altitudes as low as 140 km could be sampled. If cheap generalized detectors can be mounted in the "pearls on a string" configuration, then very exciting measurements of vertical profiles, divergence, and vorticities become possible. A downward-directed tether deployed along field lines at high latitudes would be able to investigate upwelling ion streams, auroral electrodynamics, ion acceleration regions, neutral viscosity, and non-Maxwellian flows, etc. Detailed altitude profiles of minor constituents, and state variables, and the vertical propagation of gravity waves and tides could also be studied. If the tether could be rotated in the horizontal plane along the orbital path of the mother vehicle, then horizontal gradients could be obtained. Such tether measurements would go a long way towards removing the problematical temporal/spatial ambiguity. A downward or horizontally-rotating tether mission could be flown in association with the polar platform of the space station at relatively high inclinations. Low-inclination orbits would also be of great utility for equatorial ionospheric and thermospheric work.

Technology Requirements:

- Aerodynamics and aerothermal effects on lower satellites
- Non-uniform weight distribution effects on tether

Points of Contact:

    Science Panel: Dr. Edward P. Szuszczewicz (703) 734-5516
                                Laboratory for Atmospheric and Space Sciences
                                Science Applications International Corporation
                                1710 Goodridge Drive
                                McLean, VA 22102

    Discipline: Ionospheric, thermospheric, & mesospheric physics
    Program Manager: TBD
    Program Scientist: Dr. Dave Evans (202) 453-1514
    NASA Center: MSFC
    Project Manager: Carmine DeSanctis (205) 544-0618
    Project Scientist: TBD

6.4 ENA/EUV Imager

Target: Magnetospheric plasma regions

Orbit: Lunar base

Mission Duration: >1 year

Mission Class: Space Exploration Initiative (SEI)

Mass: ~ 200 kg (instruments only)

Launch Vehicle: TBD

Theme: Magnetospheric physics

Science Objectives:

Ultraviolet/extreme ultraviolet imaging:

- Global/macroscale observations of plasmasphere, plasmasheet, magnetotail
- Derivation of global distributions and dynamics
- Observation of substorm and storm effects
- Inference of global magnetospheric configuration and dynamics
- Pointing accuracy ~0.05°
- Achievement of 10 kbps data rates

Energetic neutral atom imaging:

- Global/macroscale observations of magnetosphere for global distributions and dynamics of charged particle regions
- Observation of substorm and storm effects
- Earth-centered field of view, minimum pixel resolution of 0.5° providing spatial resolution of 0.5 RE and 10 kbps data rate

Spacecraft:

    Type: Lunar base observatory
    Special Features: TBD
    Special Requirements: TBD

Section 6 Candidate Future Missions

Instruments:

	Mass	Power	Data rate	Data storage	FOV
UV/EUV Imager	~50 kg	TBD	10 kbps	TBD	40°
ENA Imager	~50 kg	TBD	30 kbps	TBD	TBD

Mission Strategy:

Continuous observations of magnetosphere and magnetotail by neutral atom and scanning imaging instruments from lunar based observatory.

Technology Requirements:

The Plasmasphere Imager (He+, 30.4 nm) and Plasmasheet Imager (O+, 83.4 nm) will be developed. Development of improved sensitivity from lunar-based observatory may be required.

Points of Contact:

Science Panel:	Dr. Andrew F. Cheng (301) 953-5415 Applied Physics Laboratory Johns Hopkins University Johns Hopkins Road Laurel, MD 20707
Discipline:	Magnetospheric physics
Program Manager:	TBD
Program Scientist:	Dr. Tom Armstrong (202) 453-1514
NASA Center:	MSFC
Project Manager:	Carmine DeSanctis (205) 544-0618
Project Scientist:	TBD

6.5 Magnetopause Mapper (Ionosonde)

Target: Earth's magnetosphere

Orbit: Lunar base

Mission Duration: 3+ years

Mission Class: Space Exploration Initiative (SEI)

Mass: TBD

Launch Vehicle: TBD

Theme: Magnetospheric physics

Science Objectives:

Conduct active sounding of the magnetopause boundary, using a radio transmitter in 2–100 kHz range to sound the magnetopause from a lunar base (distance, structure, and motion). Preliminary assessment indicates that echoes from plasmoids approaching and leaving the lunar distances can be measured.

Primary objectives are as follows:

- Determine plasma gradients in the magnetopause boundary layer.
- Determine magnetopause boundary motions.
- Search for plasmoids propagating down the magnetotail (assuming lunar backside site).
- Determine the waveguide propagating characteristics of the magnetosphere.
- Investigate structures and dynamics of the magnetotail, including the low-latitude boundary layer and the distant plasma month.
- Determine changes in the magnetospheric tail configuration and magnetic flux caused by external and internal processes.

Spacecraft:

    Type: Lunar base facility

    Special Features: Single dipole transceiver and ~10, 2 km receiver dipoles in array (~30 km square)

    Special Requirements: Pulse/coded output

Section 6 Candidate Future Missions

Instruments:

	Mass	Power	Data rate	Data storage	FOV
2–100 kHz radio transceiver (9)	Total 9000 kg	200 W per transceiver	900 kbps	N/A	360°
Phased antenna array	TBD	TBD	(local)	10 Mb	360°
Communication/ processing center	~30 kg	TBD	2 kb	40 Mb	N/A

Mission Strategy:

Use phased array receiver (collinear dipoles) placed on the lunar surface in support of radio telescope needs or dedicated to this project. Add transmitter wire dipole for isotropic output signal. 10 ms pulses allow 0.25 R_E spatial resolution. 1 W bursts yield return signal at least 60 dB (nearside) or 90 dB (farside) over continuum background. 10 W signal preferable for nearside use.

Technology Requirements:

- Lunar base
- Deployed wires on lunar surface

Points of Contact:

Science Panel:	Dr. Patricia H. Reiff (713) 527-8750 Department of Space Science and Astronomy Rice University P.O. Box 1892 Houston, TX 77005-1892
Discipline:	Magnetospheric physics
Program Manager:	TBD
Program Scientist:	Dr. Tom Armstrong (202) 453-1514
NASA Center:	MSFC
Project Manager:	Carmine DeSanctis (205) 544-0618
Project Scientist:	TBD

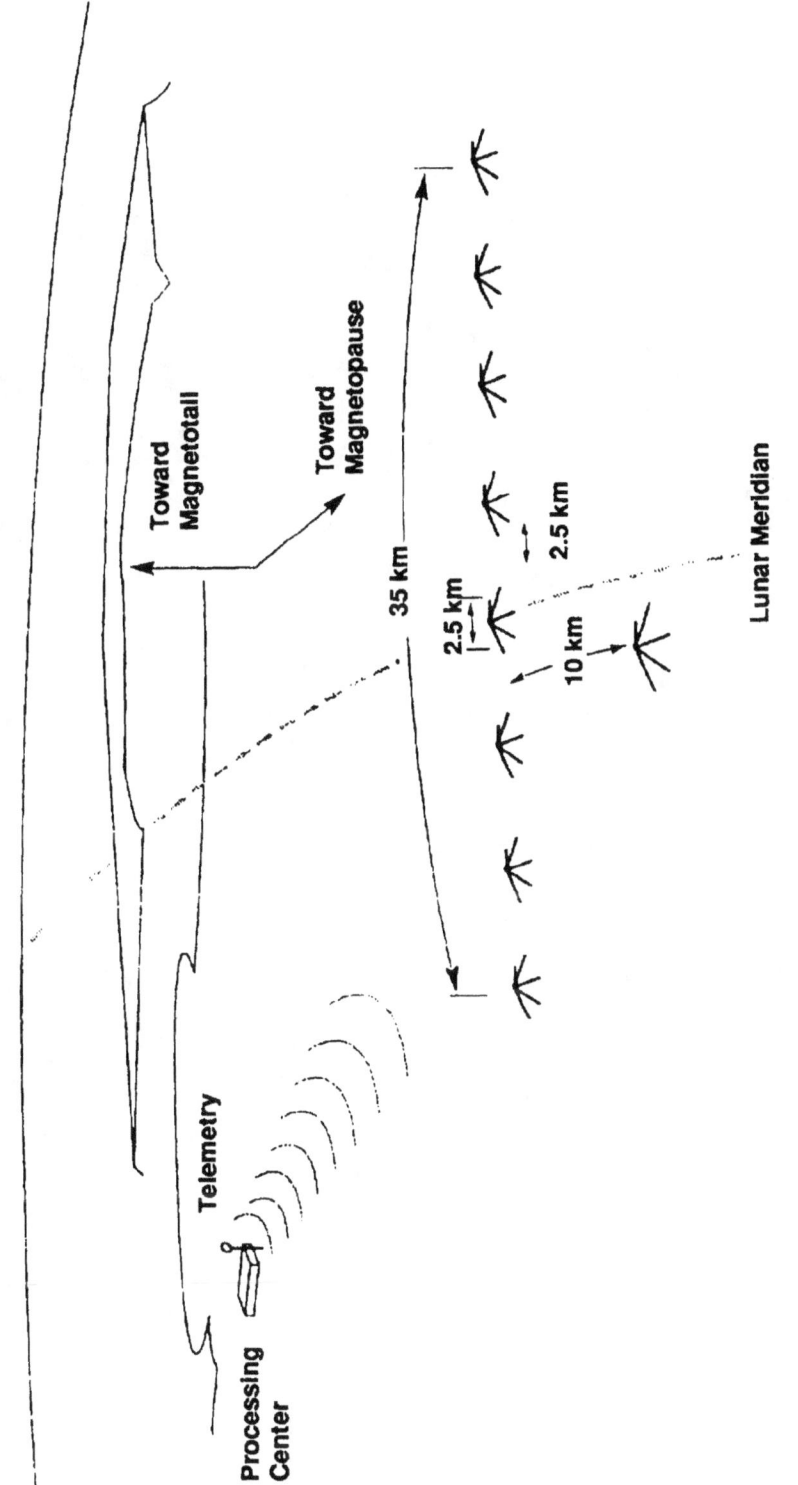

Lunar-based Magnetosphere Sounder Array concept

6.6 Lunar Solar Observatory

Target: Multi-wavelength high-resolution observations of the Sun

Orbit: Lunar base

Mission Duration: Indefinite with phased development

Mission Class: Space Exploration Initiative (SEI)

Mass: 2500 kg increasing 10,000 kg per year for a total of 100,000 kg after 15 years

Launch Vehicle: TBD

Theme: High-resolution observations of the solar atmosphere across all wavelengths from gamma rays to radio

Science Objectives:

Determine:

- The basic plasma physics processes responsible for the metastable energy storage and impulsive energy release in solar flares

- The cause of the solar activity cycle and the factors that control the structure and dynamical behavior of the solar magnetic field, the photosphere, chromosphere, and corona

- The factors that determine the MHD structure and behavior of the convection zone

- How the structure of the heliosphere changes through the action of the solar wind.

Spacecraft:

    Type: Lunar base observatory

    Special Features: Solar pointing, real time data processing, lunar base technology

    Special Requirements: Man-tended

Instruments:

	Mass	Power	Data rate	Data storage	FOV
UV/O/IR telescope	15000 kg	3 kW	20 Mbps	1014 bits	TBD
P/OF	15000 kg	3 kW	10 Mbps	TBD	TBD
SXT facility	15000 kg	3 kW	20 Mbps	TBD	TBD
HEF	15000 kg	3 kW	5 Mbps	TBD	TBD
Radio facility	10000 kg	3 kW	TBD	TBD	TBD
Support facility	30000 kg	15 kW	5 Mbps	TBD	TBD
TOTAL	100000 kg	30 kW	60 Mbps	TBD	TBD

These are estimates for the complete observatory. The early instruments would require smaller resource allotments.

Mission Strategy:

The observatory would be developed over a period of two decades with a continual upgrading of the instruments and the on-site observational and analysis center. Initially, the observatory might consist of an optical telescope designed primarily as a vector magnetograph for the flare alert system and developed by the SEI. This would be rapidly augmented by adding space-station class instruments to expand the wavelength coverage to the X-ray/XUV, UV, IR, and radio. The third stage would be to add a high-energy facility containing a lunar P/OF for hard X-ray and X-ray imaging, and various spectroscopic instruments. This would be the first of the very large facilities and should be in place by 2010 for the maximum of cycle 24. This would be followed by the long-focal-length optical telescope, upgraded coronal instruments, and the observation and analysis center.

Technology Requirements:

- Lightweight mirrors, pointing systems for lunar gravity field
- Segmented mirror technology using active systems for wavefront sensing and control
- Large area focal plane instrumentation

Points of Contact:

Science Panel:	Dr. Hugh Hudson (619) 534-4476
	CASS
	University of California at San Diego
	C-011
	La Jolla, CA 92093
Discipline:	Solar physics
Program Manager:	TBD
Program Scientist:	Dr. Dave Bohlin (202) 453-1514
NASA Center:	MSFC
Project Manager:	Carmine DeSanctis (205) 544-0618
Project Scientist:	TBD

6.7 ITM Coupler

Target: Earth's ionosphere, thermosphere, and mesosphere

Orbit:
1. 80° inclination, 120 x 4000 km (1 s/c)
2. 28° inclination, 120 x 4000 km (1 s/c)
3. 400 km circular (4 s.c)
   — Sun synchronous, 4pm ascending
   — Sun synchronous, 12 noon ascending node
   — Sun synchronous, 8am ascending
   — 45° inclination
4. Polar 1000 x 8000km (2 s/c with ascending nodes 90° apart)

Mission Duration: 6 years

Mission Class: Major

Mass:
1. 615 kg
2. 615 kg
3. 690 kg each
4. 630 kg each

Launch Vehicle: Medium ELV's (different launch sites and inclinations may require different launch vehicles)

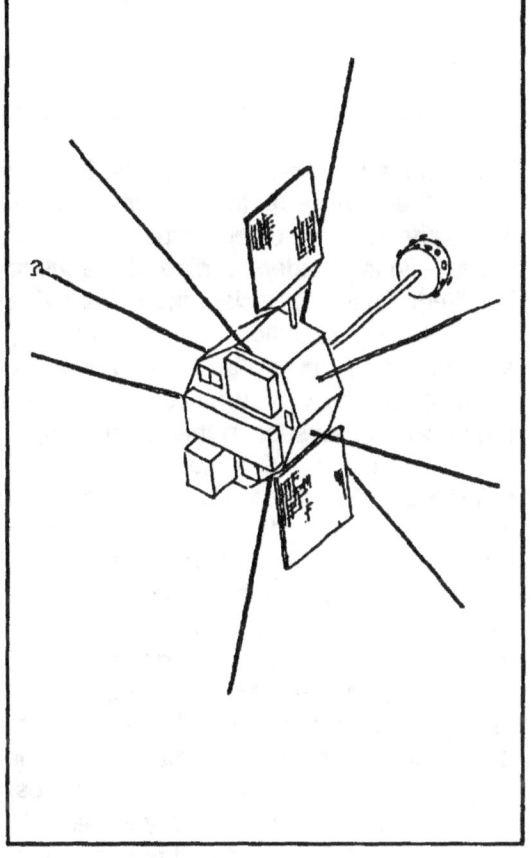

Theme: Investigation of global scale coupling and dynamics within the ionospheric-thermospheric-mesopheric system itself and to the magnetosphere and heliosphere above and the lower atmosphere below.

Science Objectives:

Determine:

- The global distributions of electric fields, thermospheric winds, and specifications of EUV/UV radiation

- The global distribution of intermediate plasma layers and their contributions to dynamo fields and specific controlling influences of metallic, atomic, and molecular ion inventory

Investigate:

- Wind shear forces and electric fields in intermediate layers
- Field line coupling of electric fields in the upper F-region

Study:

- The mesosphere/lower thermosphere interface

The Earth's Ionospheric-Thermospheric-Mesopheric (ITM) domain has a uniquely important role in the solar-terrestrial system. This geospace region provides the interface between interplanetary and lower atmospheric processes and is the most effective testbed for establishing a definitive understanding of the full spectrum of solar-terrestrial coupling processes involving plasma particles and fields and atmospheric dynamics. It absorbs the bulk of the solar EUV/UV radiation and precipitating particle energies, and supports and controls currents and potentials ranging up to 10^6 amperes and hundreds of kilovolts respectively. It is among the most complex, naturally-occurring plasma domains accessible to *in situ* diagnostics. It is a region with processes in the continuum, transitional, and free-molecular-flow regimes. It involves positive and negative ion chemistry, global circulation, kinetics of neutral particle collisions, external and internal electric fields, and magnetically-induced anisotropies. It couples to the magnetosphere and heliosphere above, and the stratosphere below—responding to solar EUV/UV variations, coronal mass ejections, high-speed streams from coronal holes, atmospheric tides, gravity waves, and lighting storms. It responds interactively with the magnetosphere, controlling and modulating currents and fields, and contributing to the major components of the magnetospheric ion population. In all cases, ITM responses to coupled internal and external forces are complex and dynamic. Its layers move up and down, its densities vary dramatically, and its composition changes in ways yet to be properly described by empirical or first-principle models. We do not know the global time-dependent distributions of electric fields and thermospheric winds nor the specifications of incident EUV/UV radiation—the primary forces driving the system. Within this system we do not know the global distribution of intermediate plasma layers, their contributions to critically-important dynamo fields, and the controlling influences of the metallic, atomic, and molecular ion inventory. In the cause-effect chain we do not understand the interplay of the wind-shear forces and electric fields in the formation of intermediate layers, their global distributions and control of upper F-region dynamics by field-line coupling of electric fields. We do not understand the physics controlling the mesosphere/lower thermosphere interface, the upward and downward propagation of tides and waves, their ultimate coupling to the ionized constituents, and the cascade of energy between large and small scale structures. It is the physics of the ITM domain and the unknowns listed above that are addressed in the mission of the ITM Coupler. The mission is one of investigation and exploration, focussing on the global, dynamic and coupled system, its role in the solar-terrestrial chain and the near-Earth environment.

The overall objective is the development of a self-consistent theoretical and empirical understanding of the ITM as a single, electrodynamic, chemically-active, and kinetically-reactive fluid. As an integral part of the objective, the understanding must be global, cover the full spectrum of solar-terrestrial conditions, deal with all internal and external coupling mechanisms, and encompass quiescent and dynamic conditions carried to the limit of highly irregular and unstable modes.

Within the framework of the overall objective the issue of "coupling" takes on a particularly important role. If we are to understand the ITM, we must understand the solar, interplanetary, magnetospheric, and atmospheric controls, We must understand mesospheric/thermospheric coupling due to gravity wave, tidal, and trace-constituent transport processes at various scales. We must understand the electrodynamic coupling between the thermosphere/ionosphere and the magnetosphere and the processes which control currents, fields, and plasma flows. We must understand the coupling processes between large- and small-scale structures and the effects of turbulence and instabilities on current systems and transport in general. And we must emphasize the need to understand the coupling between charged and neutral constituents in the ITM system, as that coupling takes place both locally and over large-scale vertical and horizontal domains.

The National Academy of Sciences points to the magnitude of this incomplete understanding of the ITM by stating that "the Earth's mesosphere and lower thermosphere and ionosphere are the least explored regions in the Earth's near-space....[Its] overall structure and dynamic responses to magnetospheric substorms, solar flares, and stratospheric warmings, and even the basic controlling physical and chemical processes of these effects are not understood....We still do not understand the basic processes that drive and control the global electric circuit and [we know that the model of] the ionosphere as a highly-conducting equipotential boundary is incorrect."

To advance our understanding of solar-terrestrial coupling processes and the uniquely important role of the Earth's ionospheric-thermospheric-mesopheric system significantly, the ITM Coupler will address the following set of fundamental questions:

- What is the relative contribution of the various energy sources (UV/EUV, auroral and ring-current particles, Joule heating, tides and gravity waves, etc.) and sinks (ionization, excitation, thermal conduction, etc.) to the overall ITM energy budget?

- How is the energy within the ITM system transferred and dissipated in transition between large and small scale phenomenologies? And how does this energy couple upward and downward into neighboring domains?

- How do global ionospheric plasma distributions and associated current systems respond to magnetic storms?

- What are the global distributions of electric fields and thermospheric winds under quiet and disturbed conditions?

- What are the relative roles of winds, electric fields, and ion composition in plasma layer formation and dynamics?

- What is the global distribution of intermediate, descending and sequential plasma layers and what are their roles in dynamo-driven electric fields?

- How does the dynamic behavior of the neutral atmosphere influence electro-dynamic coupling within the ITM and in its ties to the magnetosphere?

- What is the spectrum of dynamic behavior of gravity waves and tides in the thermosphere and upper mesosphere, and how is the associated energy dissipated?

- How important is large-scale thermospheric circulation in changing the composition and energetics of the mesosphere?

- How do global ionospheric plasma distributions and associated current systems respond to magnetic storms?

- Where and how do the magnetospheric and atmospheric currents close in the ionosphere?

- What is the role of the ionosphere in populating the magnetospheric plasma, where are the source regions, and what are the mechanisms for energetic ion outflow?

Spacecraft:

Type: 3-axis stabilized

Special Features: Capable of orbital adjustment

Special Requirements: ~300–400 m/sec dV

The spacecraft will be of two types. The low perigee orbit will require a symmetrical low drag design with minimum appendages, in the style of AE C, D&E. A more detailed Phase A study will address the effects of appendage design and the study will address the effects of appendage design and the concomitant science on spacecraft drag, Delta V, and attitude control.

The other six spacecraft will be of similar design with deployed solar arrays, three-axis stabilization and onboard propulsion.

The low perigee spacecraft will be similar in design to AE; a right circular cylinder, in concept, with the axis of symmetry perpendicular to the orbit plane. Body-mounted solar cells along with batteries will supply spacecraft power. At the beginning of life (BOL) 100 W will be available for the payload. Communication will be via S-Band to dedicated ground receiver sights. Onboard propulsion will be required to achieve low perigee and maintain apogee. Preliminary design results show a 630 kg spacecraft including a 168 kg propulsion system.

The remaining six spacecrafts will be of similar design, each about 700 kg of which 270 kg is propulsion. Payload power requirements of the 1000 x 8000 km mission will be 170 watts each while the four 400 km circular orbiting spacecraft will require 480 watts each. Communication to and from all six will be via TDRSS.

Section 6 Candidate Future Missions

Instruments:

	Mass	Power	Data rate	Data storage	FOV
Neutral & ion mass composition	TBD	TBD	TBD	TBD	TBD
Neutral, ion, electron temperature & number density	TBD	TBD	TBD	TBD	TBD
Ion drift velocity & neutral wind vectors	TBD	TBD	TBD	TBD	TBD
Electric & magnetic fields	TBD	TBD	TBD	TBD	TBD
Energetic particle environment	TBD	TBD	TBD	TBD	TBD
Trace constituent populations	TBD	TBD	TBD	TBD	TBD
Solar spectral irradiance (EUV/UV)	TBD	TBD	TBD	TBD	TBD
Global airglow emission rates	TBD	TBD	TBD	TBD	TBD

The mission objectives and associated set of scientific questions impose a requirement on the implementation plan for a multi-instrumented, multi-spacecraft configuration that includes *in situ* and remote sensing capabilities. The *in situ* capabilities will provide the highest-resolution measurements of local thermal, suprathermal and energetic particles, currents and fields, thermospheric winds, and the cascading processes between large- and small-scale features. Capabilities will include determination of Maxwellian, non-Maxwellian, or otherwise anomalous energy distributions, associated anisotropies, and cause-effect roles in momentum and energy transfer within and across ITM boundaries. These measurements will be complemented by remote sensing and imaging detectors that will determine contributing parameters at locations remote from the *in situ* track. Monochromatic imagers will monitor the exchange of chemical species, wave momentum stresses, and particle and wave energy between the mesosphere and thermosphere. Time-dependent observations of atmospheric emissions will determine energy deposition from solar radiation and particle precipitation; and remote sensing of major and minor thermospheric constituents will determine the baseline state of the upper atmosphere and monitor possible anthropogenic changes. The unique combination of *in situ* and remote sensing techniques will enable a more detailed understanding of coupling processes and expand the measurement profiles to 2 and 3 dimensions. A list of the required measurements and instrument types is included in the following table.

Section 6 Candidate Future Missions

Measurement Requirements and Associated Orbits

Primary parameters	Instrument type	Spacecraft orbit		
		400 km circular	120 x 4000 km elliptical	1000 x 8000 km elliptical
Ion and neutral composition & density	Mass spectrometer	X	X	X (Ions only)
Electron density, temperature & irregularity structure	Langmuir Probe	X	X	X
Neutral wind velocity & temperature	A. *In situ* neutral wind detector	X	X	
	B. Remote sensing Fabry-Perot interferometer	X		
	C. VIS/UV spectrophotometer (temp. & density only)		X	
Electric field	A. 3-axis double probe	X		X
	B. Ion drift meter	X	X	
Magnetic field & currents	Magnetometer	X	X	X
Thermal/suprathermal ion & electron energy distributions	Retarding potential analyzer	X	X	X
Auroral imaging & optical signatures of energy deposition	A. Monochromatic imager & scanner (EUV/UV)	X		
	B. Full disc imager			X
	C. Doppler imaging (5577 Å, 6300 Å, 7320 Å)			X
Sodium layer morphologies	NA Lidar	X		
Solar EUV/UV fluxes	EUV/UV photometers			X
Energetic particle fluxes	Particle spectrometers	X		X

Mission Strategy:

The eight spacecraft will be placed in their respective orbits by MELVs from both the ESMC and WSMC. The highly elliptical low perigee spacecraft pair could be candidates for a downscoping (during a Phase A study) enabling them to be launched on a Scout or Pegasus class launch vehicle. Communications with the spacecraft will be via TDRSS for the 1000 x 8000 km and the 400 km circular orbiting set. Dedicated ground receiver(s) will be required for the low perigee (120 km) portion of the mission.

The development of a comprehensive understanding of the ITM system implies an understanding of seasonal, solar cycle and diurnal controls, as well as the associated altitudinal, latitudinal and longitudinal responses to quiet and disturbed conditions. The scientific objective also requires that the understanding be self-consistent and treat the coupled, global-scale system. These criteria require three types of spacecraft orbits: 1) circular, low-altitude (approx. 400 km) and long-lived, to provide global distributions of electric fields, thermospheric winds, upper mesospheric structures, particle and radiation inputs, and dynamic plasma interactions, 2) low-perigee, elliptical (approx. 120 x 4000 km) to penetrate the lower thermospheric and ionospheric domains for *in situ* measurements of localized distributions of winds, electric fields, ion inventories and intermediate layer distributions, and 3) high-altitude, elliptical to provide synoptic large-scale imaging and scan-platform diagnostics of auroral oval dynamics, thermospheric structures, and ionospheric/atmospheric responses to energetic inputs. There is an attempt in this configuration of satellite orbits to keep the measurements systematically

Section 6 Candidate Future Missions

comprehensive and eliminate (wherever possible) unnecessary variables (*e.g.,* constantly changing spacecraft altitude). This configuration employs a baseline of eight spacecraft-Four in low altitude circular orbits, and two each in low orbits, and high altitude elliptical orbits. The four circular spacecraft would all be at the same altitude (=400 km). Three would be Sun-synchronous (4am/4pm, noon/midnight, and 8 am/8pm), and one would be in low-to-mid inclination orbit (45°–57.5° may be optimum). The four circularly-orbiting spacecraft would be phased so that the low-inclination orbiter routinely established simultaneous and successive cross-track coordinate registration with each of the Sun-synchronous satellites. Every 90 minutes there would be six cross-track encounters (three dayside and three nightside), providing "local" synchronized meridional and zonal data sets. Coupling the *in situ* and remote sensing techniques with coordinated ground-based diagnostics would provide extensive three-dimensional coverage with a systematically acquired database of UT, LT latitudinal, longitudinal, solar cycle, and seasonal variations. In the same 90-minute period the three Sun-synchronous spacecraft would also provide synchronized measurements of the convection electric field, currents, precipitating particles and thermospheric winds across the north/south polar regions. This approach provides comprehensive mapping of thermospheric winds, upper mesospheric structures, electric fields, ion composition (with relevance to metallic ion inventory necessary to "feed" the underlying intermediate layers), and scale-size distributions of various phenomena from 100's of kilometers to meters. The database would be unprecedented from an experimental point of view and a global-scale modelling perspective. The configuration also includes two low-altitude elliptically-orbiting spacecraft (perigee/apogee = 120km/400km) to penetrate the lower ionospheric/thermospheric domains in high and low inclination orbits, and two high-altitude elliptical orbiters. The high-altitude elliptical orbits, currently specified as 1000 x 8000 km polar are under consideration for implementation as Molniya type orbits (2000 x 37,5000 km, i=63.4°, fixed periapsis) to provide auroral and airglow images at high apogee, and Doppler wind measurements during perigee passes. The Molniya orbit study has not been completed, nor has the final instrument and spacecraft fleet configuration been defined.

The ITM Coupler telemetry will be transmitted to the Goddard Space Flight Center (GSFC) ground operations and data system via TDRSS or a designated ground station. Data transport services will be performed by the NASA Communications (NASCOM). Other ground facilities needed to support the ITM Coupler will include Payload Operations Control Center (POCC), Command Management Systems (CMS), Flight Dynamics Facility (FDF), Network Control Center (NCC), and the Data Capture Facility (DCF). The science ground system will be equipped with a centralized ITM Coupler Science Data Processing Center (ISDPC) for generation of higher level data products (level 1,2,3, etc.) and an ITM Coupler Science Operations Center (ISOC). The ITM Coupler ISOC will be used as the interface with the Principal Investigators (PI's) and Co-Investigators (Co-I's) for science planning and operations. The PI and other science users will receive science data products and ancillary data from the ISDPC for further data processing and analysis at the remote sites.

The ITM Coupler commands initiated from the operational staff will be uplinked via POCC at 1 kbps via Tracking and Data Relay Satellite System (TDRSS) Multiple Access (MA) or the designated ground station.

Ground data operations and processing elements fro the ITM Coupler will include the White Sands Ground Terminal, the designated ground station, and the mission operations and data processing facilities mentioned above. The ISOC and ISDPC will also be involved in the ground data operations system. The ITM Coupler coverage data and mesospheric/ thermospheric science data will be transmitted to the ground mission operations and data processing system via TDRSS or the designated ground station.

Telemetry data will be transmitted as required to the POCC for spacecraft housekeeping data, to the FDF for spacecraft orbit/attitude data, and to the DCF for science data. The POCC will be the mission control center, providing mission planning/scheduling, spacecraft health and safety monitoring, end-to-end testing, simulation, and analyzing customer requests such as checking with network control and TDRSS support. The CMS will receive observation requests for the spacecraft from the PI's and Co-I's, will analyze requests against spacecraft constraints, and will generate the command sequence for approved requests for delivery to POCC for uplinking.

Spacecraft data such as orbit, universal time, and attitude, will be received by the FDF through NASCOM. The POCC and CMS interact so that spacecraft commands will be uplinked to the ITM Coupler. Science data will be transmitted through NASCOM to the DCF, where data will be captured and Level Zero processed. The DCF will also provide data accounting and data storage services, and data distribution to the ISDPC. The ISDPC will perform high-level data processing on the received science data. The PI's, Co-I's, and other science users will receive science data products and ancillary data for further processing and analysis on a request basis.

Technology Requirements:

During Phase A studies the low perigee (120 km) portion of the mission would be examined with the goal of making the spacecraft compatible with lower cost launch vehicles. Additionally, the compatibility of making field measurements through appendage deployment and the resultant aerodynamic effects on the low perigee spacecraft would be studied.

Tradeoffs between high altitude elliptical orbits (1000 x 8000 km) and Molniya orbits will also be addressed.

Points of Contact

Science Panel:	Dr. Edward P. Szuszczewicz (703) 734-5516 Laboratory for Atmospheric and Space Sciences Science Applications International Corporation 1710 Goodridge Drive McLean, VA 22102
Discipline:	Ionospheric, thermospheric, & mesospheric physics
Program Manager:	TBD
Program Scientist:	Dr. Dave Evans (202) 453-1514
NASA Center:	GSFC
Project Manager:	Steve Paddack (301) 344-4879
Project Scientist:	TBD

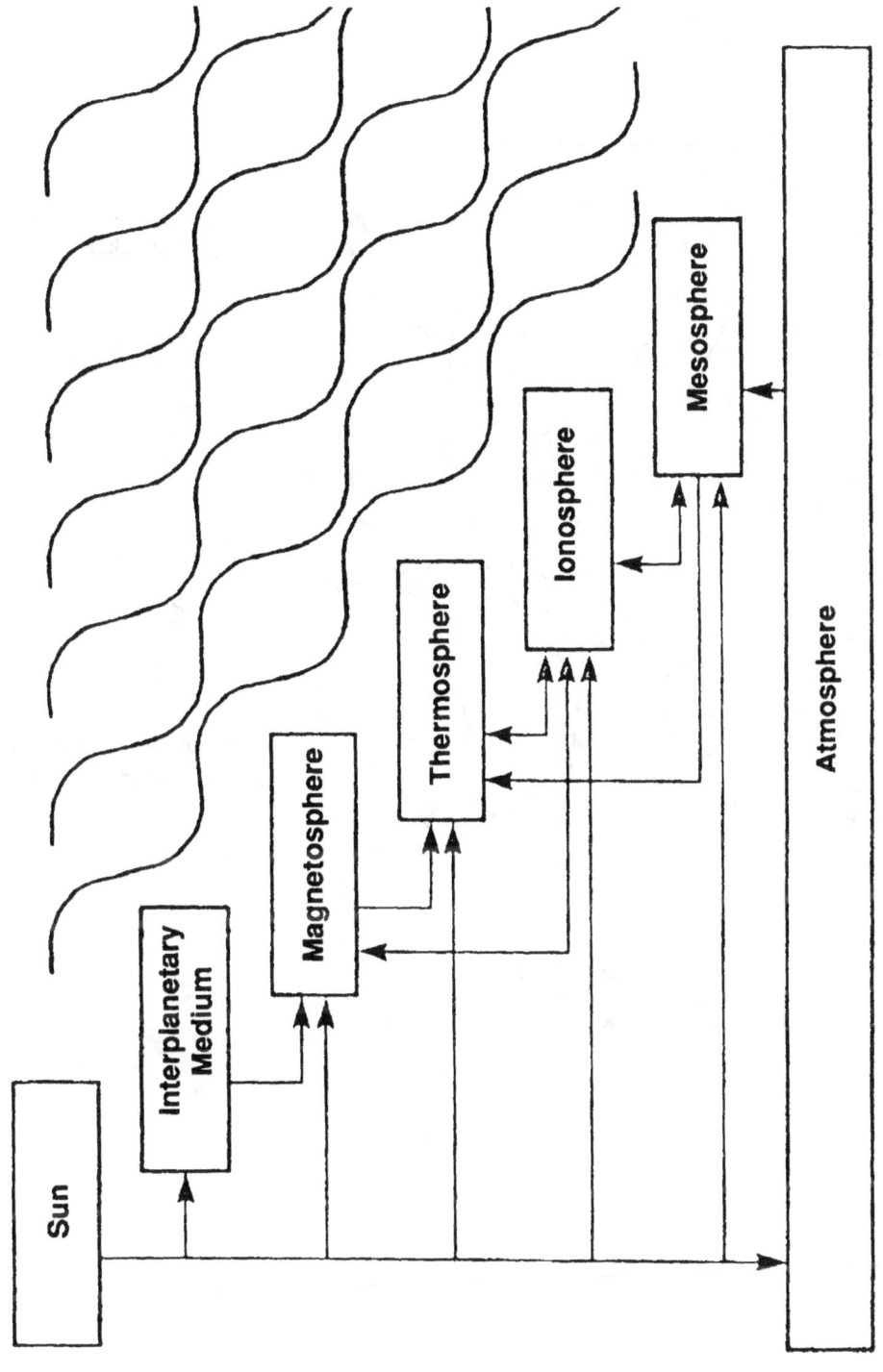

Coupling and dynamics in the cascading of particles, fields, and waves in the solar-terrestrial environment

Section 6 Candidate Future Missions

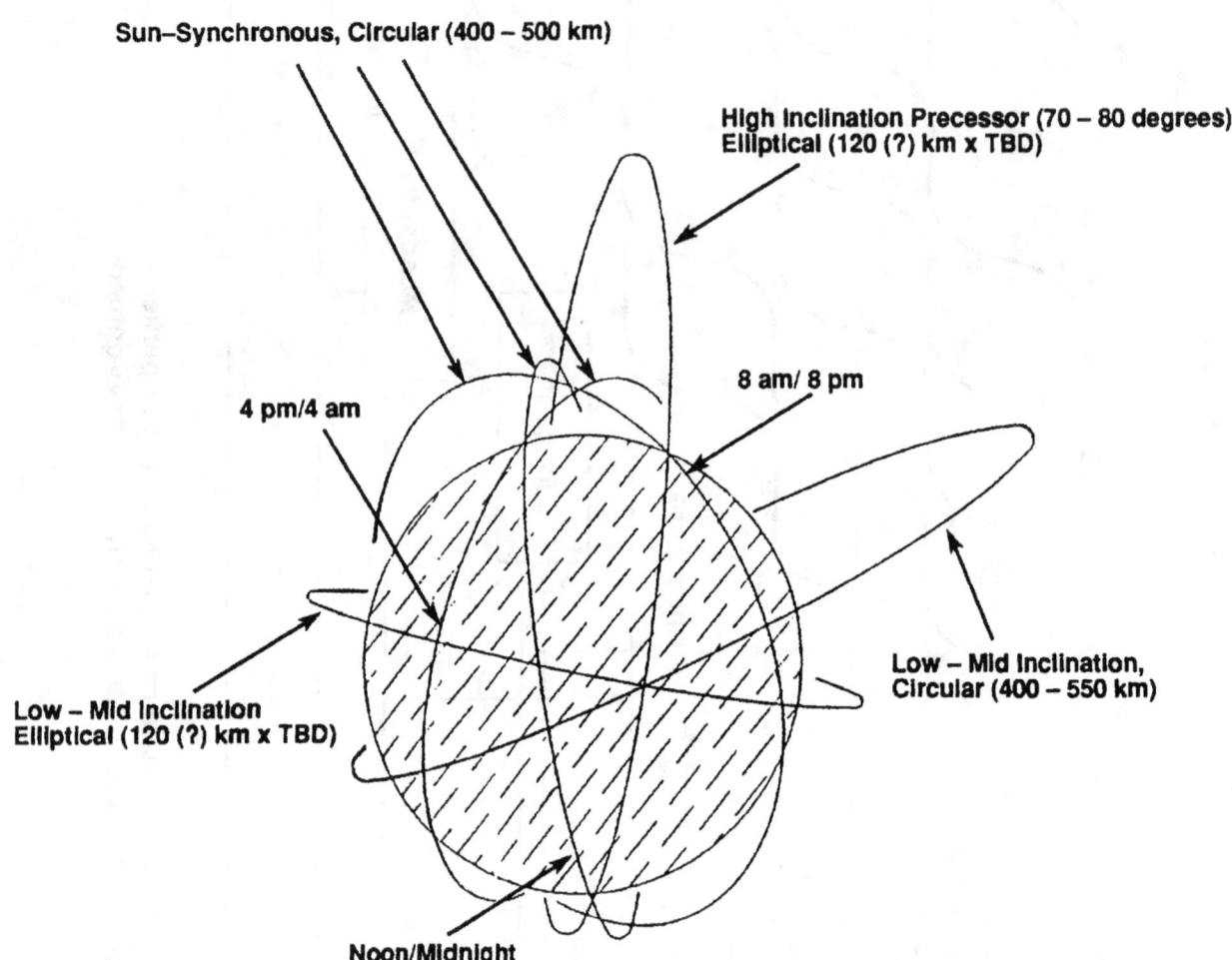

ITM Coupler orbit configuration

Section 6 Candidate Future Missions

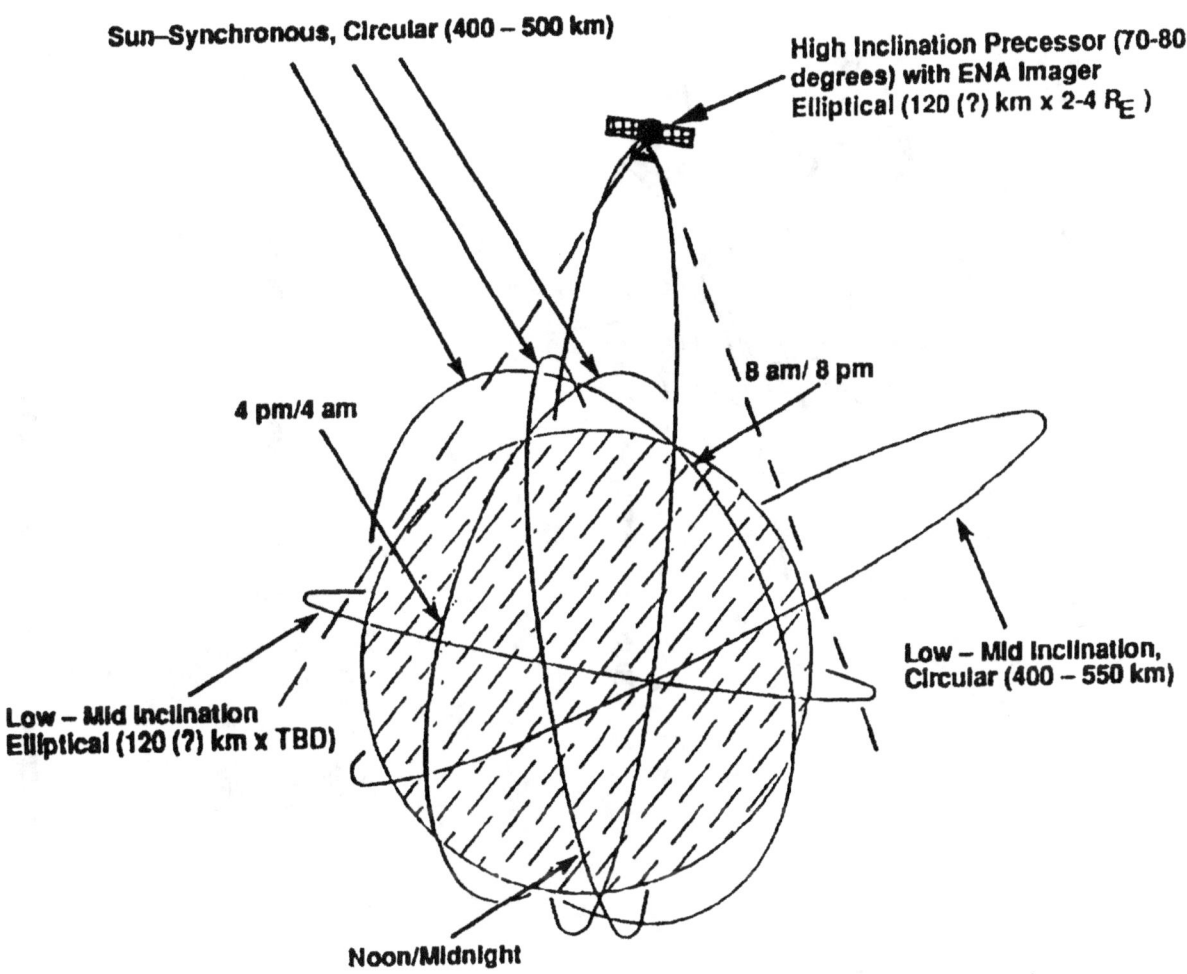

ITM Coupler/Energetic Neutral Atom (ENA) imager

Section 6 Candidate Future Missions

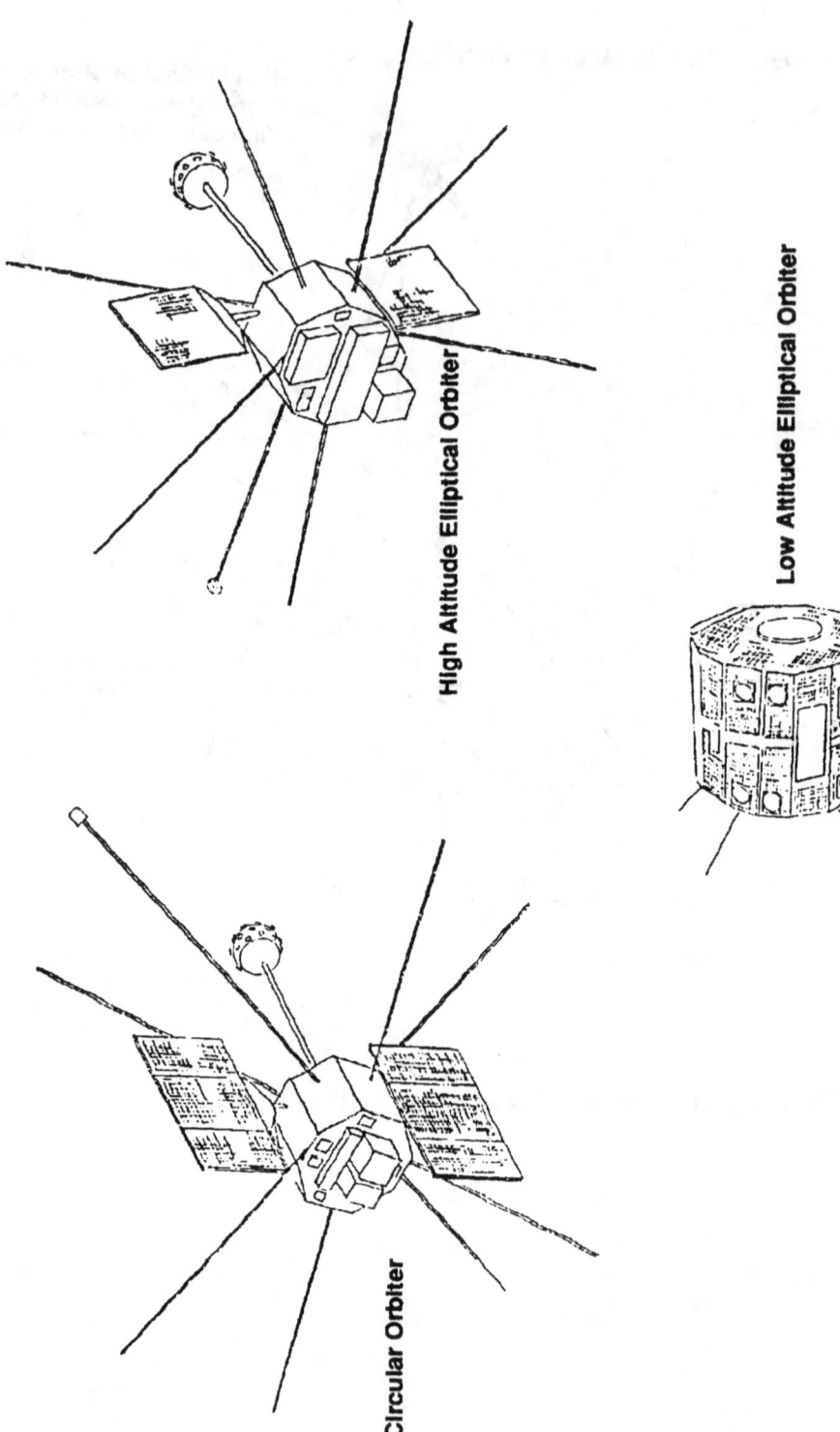

Ionosphere, Thermosphere, Mesosphere Coupler

6.7-12

6.8 Mesosphere Structure, Dynamics, & Chemistry

Target: Earth's mesosphere

Orbit: 600 km, circular, near polar, Sun-synchronous
3 pm ascending node

Mission Duration: 12 years

Mission Class: Moderate

Mass: 1350 kg

Launch Vehicle: Medium ELV

Theme: Explore and understand the mesosphere and lower thermosphere/ionosphere

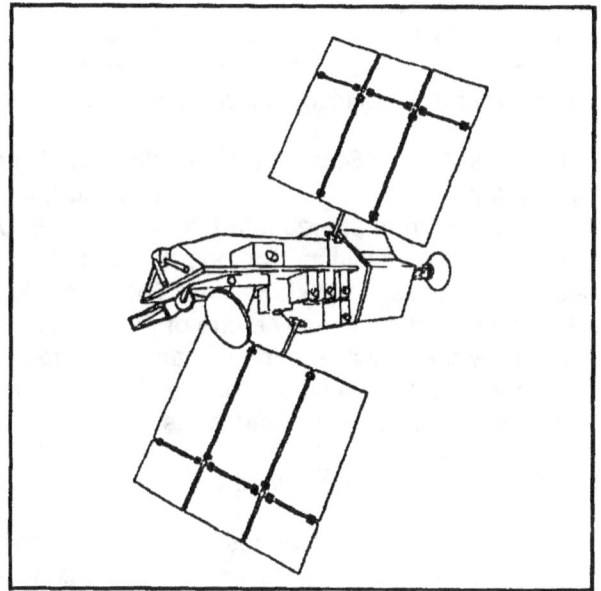

Science Objectives:

The mesosphere, lying between about 50 and 95 km altitude, is an atmospheric region with complex dynamics and photochemistry that is highly susceptible to changes in solar EUV/UV radiation. It is a region that is too high for direct sampling from balloon-borne instrumentation, yet too low for *in situ* measurements from orbiting spacecraft. In this region, very large departures from radiative equilibrium occur that are driven in part by vigorous mean meridional circulation, with rapid upward motion to cool the summer mesopause and downward motion to warm the winter mesopause by adiabatic expansion and compression, respectively. These motions are related to significant transports of chemical species that also act to change the radiative/dynamic environment. In the mesosphere, gravity waves and tides, excited both by *in situ* processes and by processes at lower altitudes, are dissipated, producing large amounts of heat and momentum that have a fundamental influence on the atmospheric structure and energetics.

The mesosphere is one of the least well understood regions of the Earth's atmosphere, but one which is susceptible to significant short and medium term changes due to anthropogenic activity. Specifically, the terrestrial mesosphere will experience anthropogenically-caused, first-order changes in its chemistry, composition, temperature and dynamics within the next 2–3 decades. Mesospheric ice clouds will spread equatorward from the summer poles, influencing the climate down to the surface. The lower ionosphere will be perturbed by larger numbers of water cluster ions. These large and profound changes are due to the increasing levels of methane and CO_2 in the middle atmosphere. Methane, which is increasing in abundance by 1% to 2% per year, is readily oxidized into water vapor. The consequent increase in mesospheric H_2O drives a stronger HO_x chemistry with far reaching effects on ozone and its heating within the mesosphere. In the complex interplay among chemistry, composition,

Section 6 Candidate Future Missions

radiation, atmospheric structure and dynamics, virtually every aspect of this region of the atmosphere will undergo change. These changes will project their influence upwards into the thermosphere and ionosphere in undetermined ways. The mesosphere is therefore an important test bed for studies of global change because of its high susceptibility to changes induced by varying abundances of trace constituents such as CO_2 and methane. There is an urgent need to establish the current mean state of the mesosphere so that we can investigate future anthropogenically-caused perturbations to that state.

This Mesosphere Structure, Dynamics and Chemistry (MSDC) mission will be designed to investigate the photochemistry, composition, dynamics, radiative heating and cooling of the mesosphere in unprecedented detail using IR, UV, and visible remote sensing techniques. The mission will be designed for long duration (11+ years) in order to determine the long-term trends in mesospheric chemistry. These trends are of critical importance since it is expected that the mesosphere will be one of the first upper atmospheric regions to demonstrate unambiguously the effects of global change due to anthropogenic causes. The instrumentation will measure factors involved in ozone chemistry, various chemical families, noctilucent clouds, mesopause temperature structures, etc.

The specific mission science questions include the following:

1. What long-term trends in mesospheric structures occur due to anthropogenically-induced chemical changes associated with increased CO_2 and methane levels?

2. What is the basic compositional, energetic, radiative, and dynamical structure of the mesosphere?

3. What are the mesospheric consequences of relativistic electron precipitation and proton precipitation events?

4. What dynamical and chemical controls are responsible for the summer mesopause anomaly?

5. What are the controlling interactions among the photochemistry, dynamics, and radiative transfer in the mesosphere?

6. What is the general circulation of the mesosphere?

7. What role do gravity waves play in mesospheric winds, temperatures, and composition?

8. What are the characteristics of the HO_x, NO_x, and O_x chemical cycles and their response to solar radiative inputs, particle precipitation, and disturbances from the lower atmosphere?

9. What is the role of Non-Local Thermodynamic Equilibrium (NLTE) processes as cooling and heating mechanisms?

The measurements to be made that address these questions are given in the following table:

Geophysical parameter	Precision %/year	Altitude region (km)
State variables		
Density	5	10–140
Temperature	1	10–140
Pressure	5	10–140
Winds	3 M/S	10–140
Composition		
Methane CH_4	0.5	10–120
Water H_2O	0.5	10–100
Ozone O_3	0.5	10–120
Carbon monoxide CO	0.5	10–140
Hydroxel OH	0.5	10–100
Atomic oxygen O	0.5	10–100
Oxygen O_2	0.5	10–120
Carbon dioxide CO_2	0.5	10–140
Nitric oxide NO	0.5	10–140
Nitrogen dioxide NO_2	0.5	
Chlorine radical C10	0.5	
Hydrogen radical HO_2	0.5	
Imaging waves	Horizontal scale (10x10 km)	40–140

Spacecraft:

    Type: 3-axis stabilized

    Special Features: Fine pointing

    Special Requirements: Correlative radar and rocket overflight programs

Section 6 Candidate Future Missions

Instruments:

	Mass	Power	Data rate(bps)	Data storage	FOV
Microwave Limb Sounder	TBD	TBD	3000	*	TBD
Infrared Limb Sounder	TBD	TBD	3000	*	TBD
Near IR Spectrometer	TBD	TBD	256	*	TBD
Near IR Sounder	TBD	TBD	256	*	TBD
Nadir IR Sounder	TBD	TBD	8000	*	TBD
Wide Angle Michelson	TBD	TBD	2000	*	TBD
Fabry-Perot Interferometer	TBD	TBD	5000	*	TBD
Imaging Fast Filter Photometer	TBD	TBD	8000	*	TBD
Ultra-Violet Spectrometer	TBD	TBD	128	*	TBD
Ultra-Violet Imager	TBD	TBD	8000	*	TBD
Na Lidar	TBD	TBD	256	*	TBD
HF Radar	TBD	TBD	TBD	*	TBD
TOTAL			43,000 bps		

*The onboard data storage system will be able to store up to 2 orbits of data

Mission Strategy:

The mission would comprise a single spacecraft in a near-polar, Sun-synchronous orbit, over one solar cycle. The spacecraft will be placed in a 600 km, circular, Sun-synchronous 3 pm ascending node orbit by MELV launched from the Western Space and Missile Center (WSMC). The spacecraft will be stabilized in three axes and will have onboard propulsion to correct orbit injection errors and to make up for drag losses during the life of the spacecraft. Deployed solar arrays and batteries will furnish power for the spacecraft and instrument operation. TDRSS and a Project Operations Control Center (POCC) will be utilized for control of the spacecraft and retrieval of scientific data.

The MSDC spacecraft Communication and Data Handling (C&DH) system will store pre-planned command sequences for operations. Engineering data from the spacecraft is estimated to be 10% of science data. The packetized science and engineering data will be dumped once per orbit when the spacecraft is in TDRSS view.

All normal operations are to be supported through the Tracking and Data Relay Satellite System (TDRSS). The Deep Space Network (DSN) will be used for contingency operations

Section 6 Candidate Future Missions

when the TDRSS is not available. The packetized telemetry will be transmitted to the Goddard Space Flight Center (GSFC) ground operations and data system via TDRSS or DSN. Data transport services will be performed by the NASA Communications (NASCOM). Other ground facilities needed to support the MSDC include Payload Operations Control Center (POCC), Command Management System (CMS), Flight Dynamics Facility (FDF), Network Control Center (NCC), and the Data Capture Facility (DCF). The science ground system will be equipped with an MSDC centralized Science Operations Center (MSOC). The MSOC will be used as the interface with the Principal Investigator (PI's), Co-Investigators (Co-I's), and other science users for science planning and observations. The PI and other users will receive science data products and ancillary data from the MSDPC for further data processing and analysis at the remote sites.

The MSDC commands, initiated form the operational staff, will be uplinked via POCC at 1 kbps via TDRSS Multiple Access (MA) or DSN. No real-time interactive operations (telescience) are required.

Ground data elements for the MSDC include the White Sands Ground Terminal, and the mission operations and data processing facilities mentioned above. The MSOC and MSDPC are also involved in the ground data operations system. The MSDC packetized telemetry will be transmitted to the ground through the TDRSS or DSN. MSDC data transport services is provided by NASCOM which interacts with the POCC and NCC. The NCC enables MSDC users to access TDRSS and DSN services and ensures the quality of services.

Telemetry data is transmitted as required to the POCC for spacecraft housekeeping data, to the FDF for spacecraft orbit/attitude data, and to the DCF for science data. The POCC is the mission control center, providing mission planning/scheduling, spacecraft and instruments health and safety monitoring, end-to-end testing, simulation, and analysis of customer requests for such services as checking with network control and TDRSS support. The CMS receives observation requests for the spacecraft by the PI's, Co-I's, and other science users, analyzes requests against spacecraft constraints, and generates the command sequence for approved requests for delivery to POCC for uplinking.

Spacecraft orbit and attitude data is received by the FDF through NASCOM. The POCC and CMS interact so that spacecraft commands may be uplinked to the MSDC instruments. Science data packets are transmitted through NASCOM to the DCF, where data is captured and Level Zero processed. The DCF also provides data accounting and data storage services, and data distribution to the MSDPC. The MSDPC performs high level data processing on the received science data. The PI's, Co-I's, and other science users will receive science data products and ancillary data from the MSDPC for further data processing and analysis at the remote sites, on a request basis.

Technology Requirements:

- Precision calibrations
- Laser for the Lidar

Section 6 Candidate Future Missions

The spacecraft technology associated with the Mesospheric SDC mission is well understood; however, spacecraft and instrument operations over a full solar cycle deserve more study. During Phase A, reliability tradeoffs should be performed to determine the optimum number of spacecraft to provide accurate data over the mission's lifetime.

Points of Contact:

Science Panel:	Dr. Tim L. Killeen (313) 747-3435 Space Physics Research Laboratory University of Michigan 2455 Hayward Street Ann Arbor, MI 48109-2143
Discipline:	Ionospheric, thermospheric, and mesospheric physics
Program Manager:	TBD
Program Scientist:	Dr. Dave Evans (202) 453-1514
NASA Center:	GSFC
Project Manager:	Steve Paddack (301) 344-4879
Project Scientist:	TBD

Section 6 Candidate Future Missions

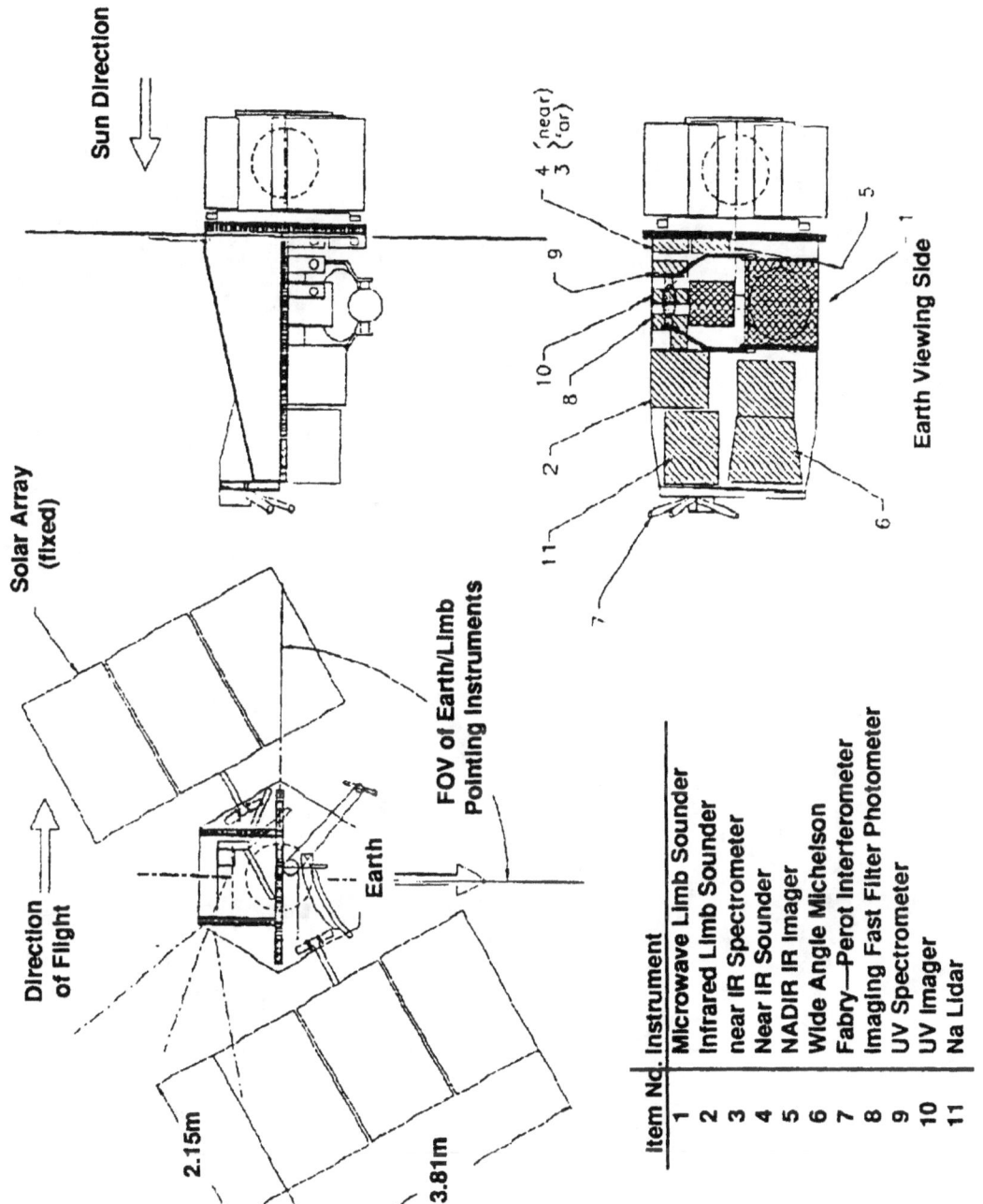

Mesosphere Structure, Dynamics, and Chemistry

6.8-7

6.9 Imaging Super Cluster

Target: Outer magnetosphere & deep tail

Orbit: Imager #1: polar, elliptical, apogee: 10–30 R_E
Imager #2: equatorial, elliptical, apogee: 10–30 R_E
4/S/C cluster: double lunar swingby, tail tour, out to ~250 R_E

Mission Duration: ≥2 years

Mission Class: Major

Mass: 2 @ 1500 kg each
4 @ 500 kg each

Launch Vehicle: 3 Atlas II launches

Theme: Magnetospheric physics

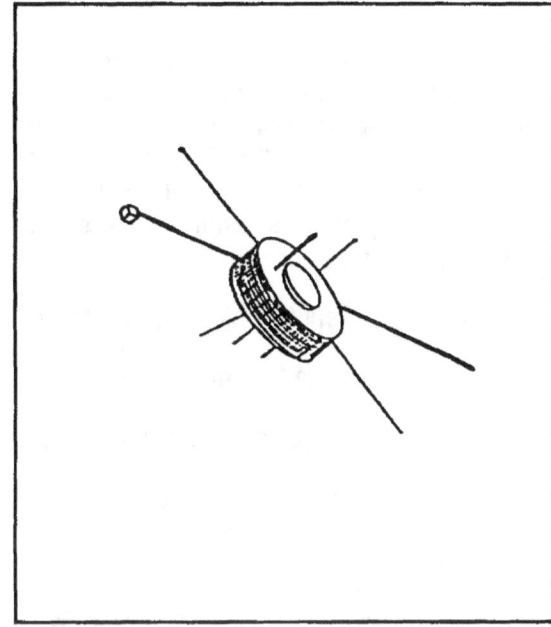

Science Objectives:

Our present understanding of the structure and dynamics of the magnetosphere has been obtained largely from *in situ* satellite measurements. With these data, the principal magnetospheric regions have been identified, and numerous plasma physical processes have been described whereby plasmas of various magnetospheric regions interact with each other and with the ionosphere and solar wind. However, it has not been possible to obtain a global understanding of how various magnetospheric regions fit together as interacting parts of a whole. While a global picture of the magnetosphere can be pieced together from numerous *in situ* measurements on a statistical basis over time, the resulting mosaic need not correspond to any of the instantaneous magnetospheric states that were sampled. It is especially difficult to assemble a coherent, unambiguous picture of global dynamical processes like substorms from a series of *in situ* measurements made by small numbers of spacecraft.

The ISC mission will obtain a combination of *in situ*, cluster-type measurements and global imaging measurements that will enable development of a coherent picture of global dynamical processes, particularly substorms, in which the entire magnetosphere is engaged in a cycle of expansion, convulsion, and contraction. The clustered *in situ* measurements will identify and characterize the microphysical processes occurring in the various magnetospheric regions, while at the same time magnetospheric imaging provides the global context for *in situ* measurements. These observations will combine synergistically to yield a greater science return than would either imaging or *in situ* measurements alone. The combination of measurements will enable temporal and spatial variations to be distinguished from each other and will allow relationships between small- and large-scale processes to be inferred. The idea is analogous to studying the functioning of an organism both by taking a movie of it and by analyzing microscopic samples of its tissues.

For a comprehensive understanding of global magnetospheric dynamics, and especially substorms, it is necessary to study the entire magnetospheric system, meaning not only the aurora, plasmasphere, and ring current, but also the outer magnetosphere and distant plasma sheet out to at least a few tens of R_E. While the aurora, plasmaspheres, and ring current plasma populations will have been imaged by various techniques on small and moderate missions prior to ISC, the distant plasma sheet will likely remain beyond their reach. Yet a clearer picture of plasma sheet dynamics on a global scale is essential for fundamental understanding of the substorm cycle and other consequences of solar wind energy input to the magnetosphere.

The principal scientific objectives for the ISC mission include the following:

- Elucidate the fundamental nature of substorms.

- Confirm the existence of plasmoids and define their role in substorms.

- Identify and characterize sources of substorm energetic particles and determine the acceleration mechanisms.

- Study plasma conditions interior and exterior to the magnetosphere that are associated with triggering of substorms.

- Define the role of ionospheric plasmas in substorms.

- Elucidate the nature of plasma sheet dynamics and clarify whether it thins, flaps, waves, and/or breaks up.

- Map the auroral and cusp boundaries as a function of time.

- Map terrestrial radio emissions as a function of time and determine their utility as a remote diagnostic of magnetospheric processes.

Spacecraft:

    Type: 2 imaging platforms; 1 cluster of 4 spinners

    Special Features: Inter-spacecraft communications; despun platform on imagers

    Special Requirements: Inter-spacecraft position determination to 1% and attitude to 0.01° for the cluster.

The spacecraft required for the mission will be of two types. The two imaging spacecraft will be three-axis stabilized with deployed solar arrays serving as a platform for the imagers. In addition to the imagers the spacecraft will carry deployable antenna for the topside sounder.

Section 6 Candidate Future Missions

The remaining four spacecraft will be spin stabilized with body mounted solar cells. The ISTP WIND spacecraft, with the deletion of gamma-ray detectors and the addition of an electric field instrument can be thought of as an analog of each of the spinners.

Instruments:

	Mass	Power	Data rate	Data storage	FOV
Imager S/C instruments: (O+ and He+ resonantly scattered radiation imaging, X-ray imaging, ENA cameras)	TBD	TBD	50 kbps nominal, 1 Mbps burst	TBD	TBD
Cluster S/C instruments: (B field to 10 kHz, E field to 1 MHz, ion composition from few eV to 50 keV, 3D electrons from few eV to 50 keV, energetic particles/electrons to 1 MeV, ion composition to few MeV/nucleon, cold plasma density, temperature, & flow velocities)	TBD	TBD	10 kbps nominal, 1 Mbps burst	TBD	TBD

The wavelengths of observation for the ISC mission will range from radio to X-ray. A number of the instruments proposed for this mission are within the present state of the art and have been flown. These include auroral, visible, UV and X-ray imagers. Also, radio receivers operating up to ~1 MHz for interferometer studies of radio emission source regions (both auroral and perhaps other acceleration regions such as plasmoids) would be used on both imaging platforms to both locate and size emission regions at kilometric wavelengths.

In addition to the above instruments of proven flight heritage, three other types of instruments are proposed which will employ new remote sensing methods. These are:

- Magnetospheric direct soundings which will use the passive radio receivers and will be used to map out magnetospheric boundaries and could determine cold plasma densities remotely.

- Energetic Neutral Atom Imagers which will use charge exchange processes in the Earth's atmosphere and geocorona to construct a remote map of the energetic ion distributions throughout the magnetosphere.

- EUV images which will map out large-scale structures of the Earth's magnetosphere such as the magnetotail and plasmoids.

Mission Strategy:

The ISC mission will employ global imaging of ENA and EUV emissions from the plasma sheet and *in situ* observations of physical processes within the tail to address the mission objectives. The ENA imager detects escaping energetic neutral atoms from charge exchange between fast ions and ambient neutral gas atoms which are mainly hydrogen atom of the geocorona. The geocoronal H density decreases from ~10 cm^{-3} near 12 R_e towards the local interplanetary value of ~0.1 cm^{-3}. The nominal desired sensitivity of the ENA imager would be ~(1–2)x10^{-2} cm^{-1} s^{-1} sr^{-1} keV^{-1} of hydrogen near 40 keV, with angular resolution of ~2° over a 40°–50° field of view achieved in ~10 minute exposures. Energy spectra will also be measured, with $\Delta E/E \leq 1$. Such an instrument would be able to obtain global images of the outer magnetosphere, with adequate sensitivity to freeze the dynamics, and with adequate spatial resolution of $\leq R_E$ to distinguish sources and acceleration processes for energetic particles. These measurements are feasible within current technology.

In addition to these imaging studies, the ENA imager will simultaneously determine the composition of the charge exchange neutrals, so as to obtain remote sensing maps of both the intensity and the composition of the plasma sheet energetic particles. The ENA imager should be able to separate atoms of hydrogen, helium, and CNO above ~keV/nuc, using thin foil time-of-flight techniques within the existing state-of-the-art. The composition information is critical to identifying the sources of plasmas in the outer magnetosphere; the helium and CNO heavy atoms detected as neutrals by the ENA imager were originally singly charged ions principally of ionospheric origin.

The EUV imager detects resonantly scattered solar ultraviolet photons, and it will use mainly He+ 304 Å and 0+ 834 Å emissions. Desired sensitivities to image the plasma sheet out to a few tens of R_E would be ~10m R (R-Rayleigh), with fields-of-view, angular resolution, and exposure times similar to those for the ENA imager. The intrinsic field-of-view for the EUV imager may be somewhat smaller than for the ENA imager, and the EUV imager may need a separate scan platform. An H Lyman α photometer/imager capability would also be highly desirable to determine H column densities along the same lines-of-sight that are viewed by the ENA imager. The EUV imager may require new technology development, such as multilayer mirrors.

The EUV and ENA imagers yield complementary measurements. The EUV imager measures total density of He+ and O+ ions, and this density is dominated by low-energy ions below several keV. On the other hand, the ENA imager measures energetic ions above several keV energy. These more energetic ions make an important and sometimes dominant contribution to the plasma pressure. In addition, the composition, of the distant plasma sheet is generally dominated by protons, which can be imaged only by the ENA technique.

Both EUV and ENA imagers measure line-of-integrals of their respective emissions. Hence, the emissions from different magnetospheric regions can be superposed on the same line-of-sight, and their contributions will need to be separated. This will be achieved by using a pair of observing platforms, one in polar orbit and one in equatorial orbit, to perform stereoscopic viewing. Each of the two ISC observatories would include a full complement of ENA and EUV imagers, and by viewing the magnetosphere from different perspectives, they will be able to distinguish emissions from different magnetospheric regions. In addition, the *in situ* cluster measurements will provide local observations in specific regions to aid in analysis of the imager.

At radio wavelengths, remote sensing will require active and passive capabilities. The active capability will involve the Magnetospheric Duct Sounder which will be essential in monitoring critical boundary layers and region interfaces which play a major role in magnetospheric dynamics. Magnetic field-aligned electron density N_e irregularities can act as wave guides for electromagnetic wave propagation. Such propagation is referred to as wave ducting. They are the result of field-aligned N_e enhancements. At higher frequencies, satellite-borne ionospheric sounders commonly observe ducted echoes of ordinary and extraordinary mode waves. In this case the ducting is due to N_e depletions. (Ducted z-mode echoes are also observed—although with less frequency.)

The success in interpreting the topside sounder ducted echoes stimulated the concept of using such a technique from a magnetospheric platform. The major feature of such a technique is the capability for remote measurements, including measurements of field-aligned NE distributions (the value of such remote measurements has been illustrated using ionospheric topside sounder data collected when the spacecraft was immersed in an equatorial plasma bubble). Such observations would provide a valuable complement to particle observations in the investigation of ionospheric ions populating the magnetosphere since the connecting path for charged particle motion is along the magnetic field. The duct-sounding technique would also enable remote sensing of large-scale features of the magnetosphere revealing structures near the cusp and the auroral zones with 0.1 R_E resolution and characterizing dayside compressions and nightside extensions of the Earth's magnetic field as proposed by Calvert [JGR, 86, 1609, 1981]. For further details, see this paper by Calvert which contained the following two tables giving pertinent parameters of the duct sounder and the ducted echoes.

Wave Source

Parameter	Value
Frequency range	20–200 kHz
Antenna length	TBD
Voltage, rms	1 kV
Current, rms	100 ma
Radiated power	1–100 mW
Pulse duration	5 ms (750 km)

Ducted Echoes

Parameter	Value
Duct width	100 km
Density variation	10% depletion
Average spacing	1000 km (10 min)
Wave mode involved	Extraordinary
Ducted power	5 µW
Echo spectral flux	10^{-18} Wm^{-2} Hz^{-1} S/N ratio (excl. AKR) 10–30 dB

Measurements Required:

Imagers

- Energetic neutral atom imaging
- Visible, ultraviolet and X-ray imaging
- Extreme ultraviolet imaging (O$^+$ and He$^+$ resonantly scattered radiation
- Duct sounding

Spinners

- Magnetic field to 10 kHz
- Electric field to 1 MHz
- 3-D electrons, few eV to 40 keV
- Ion composition to few MeV/nucleon
- Cold plasma density, temperature, and flow velocities

Using propulsive maneuvers, the four spinning satellites will be actively positioned relative to each other so as to measure the time varying, three-dimensional magnetosphere. The requirement is to know this separation distance to an accuracy of 1% or 100 meters, whichever is larger. An inter-satellite tracking system will be included on each spacecraft to determine these relative positions for separations less than about 1 R$_E$. Normal ground-based tracking will do the determination when the separations are larger.

Section 6 Candidate Future Missions

The selection of an inter-satellite tracking system is based on the fact that at the shortest separations, this measurement is virtually impossible to perform using ground-based tracking satellites and the ground station. Also, it is desirable that these measurements be made on a continuous basis. For some of the lower-altitude segments, limited viewing opportunities from the three DSN stations make this impossible. Even when stations can see the Cluster, it is not feasible to dedicate so much time on the DSN to this function. These reasons lead to the use of radio frequency tracking by the satellites themselves.

Another requirement is that when any one of the satellites detects the beginning of a scientifically interesting event it must send a trigger signal to the other satellites of the cluster. The trigger signal would change the clusters' mode of operation. The time between detection of the event by one satellite and triggering all the other satellites to the desired mode must be less than one second and, as a goal, less than 0.1 second. The inter-satellite tracking link will also serve to send this trigger signal.

To minimize the amount of extra equipment each satellite carries to perform the inter-satellite tracking function, maximum use will be made of the S-band Deep Space Network (DSN) tracking transponder that is required for communications and tracking. The raw tracking information will be collected by the satellites and the calculations of the inter-satellite separations will be made on the ground.

One implementation approach is as follows: Each satellite would be equipped with a low power S-band transmitter that is set to the same frequency as the S-band uplink from the DSN. The Cluster would be programmed so that each satellite in its turn would broadcast a 3 Mbps pseudo-noise (PN) code sequence over this link. Each satellite would use the same PN code, but with a different phase shift so that all the PN codes would be uncorrelated with each other. A low-rate data channel for satellite identification and the trigger signal is also sent.

The three satellites that are not transmitting lock up on the satellite that is transmitting and each measures the relative time (phase) between its own internally generated PN code and the received PN code. Since each satellite knows when to expect a given satellite to broadcast and what the relative phase measurements were for the last few contacts, it will be able to rapidly lock on to the received PN code. It is expected that a transmission of less than 0.1 second duration repeated once a minute by each satellite in the Cluster will be sufficient. When the is closely spread, longer transmissions may be required at the longest separations. When a satellite is performing a propulsive maneuver, it may be scheduled to increase the duration and/or the rate of its transmissions as necessary.

If a satellite determines that a scientifically interesting event has occurred, it would simply turn on its inter-satellite transmitter and start to send the trigger event message as a low data rate word along with the normal identification message. (If it were already transmitting, it would simply change the word's contents to indicate a trigger event had occurred.) The exact signal structure for this trigger function is to be determined, but it must be secure and not subject to false triggering. In other words, it should incorporate the same safeguards as a command.

When not transmitting a tracking message, and when not tracking a scheduled transmission from another satellite, each receiver will be looking for a trigger signal from any of the satellites in the cluster or a command transmission from the ground. When a trigger message is received and verified to be correct, the satellite will execute the preprogrammed mode change.

Section 6 Candidate Future Missions

When a signal from the ground is received, each satellite will determine if it is addressed to itself. If yes, it will not transmit its inter-satellite tracking signal as scheduled, but will execute whatever functions the command message tells it to do. If the message is for one of the other satellites, it will ignore the command message and continue with the inter-satellite tracking transmissions on the usual schedule.

Whenever it is neither transmitting nor receiving tracking messages, it will alternately look for a trigger signal from another satellite and a ground transmission addressed to itself. Since the ground transmission will also use a PN code, each satellite detecting the signal will note the phase difference between the received PN code and its own PN code. This measurement will be made even if the transmission were addressed to some other satellite. The ground command message and the inter-satellite tracking signals will not interfere with each other because of the different PN code correlations between them.

Each satellite's telemetry link to the ground will normally use its own PN code. The ground station will note the time difference between this PN code and its own PN code. Since the PN codes from the satellite are uncorrelated with each other, it will be possible to receive simultaneously telemetry data from all the satellites that are within the receiving antenna beam, for many of the spacing and distances all four satellites can be serviced at the same time. The satellite will also be commanded to retransmit or echo the uplink PN code. The time difference measured on the ground will then represent a two-way range measurement.

Note that the DSN does not now support PN-coded telemetry or command systems, but has expressed interest in using these techniques in the future. There is also the possibility that S-band will be phased out because of terrestrial interference. A PN modulated system may be able to combat this interference and so extend the life of S-band communications. If S-band is not available when the GTC is to fly, the other DSN frequencies will be used.

Since the tracking technique described is similar to the 30- to 100-meter accuracy demonstrated by the C-code used in the Global Positioning System (GPS), it is expected that the inter-satellite accuracy will also achieve 100 meter or better accuracy.

Because of the over-determination of the inter-satellite separations and the ranges to the ground station, one can solve for the various oscillator drifts and as it is not necessary to use the atomic clocks flown on the GPS satellites. Only reasonably good crystal oscillators are required.

It is also important to know the orientation of the Cluster relative to inertial space. This will be determined using all the inter-satellite and ground tracking data and orbit mechanics. A given ground measurement will directly measure two angles of the Cluster's orientation relative to the line of sight between the station and the Cluster. It will not directly measure the third angle which is the orientation of the Cluster "around" the line of sight. When the relative orientation between the Cluster and the station changes by a large angle, then the missing orientation angle may be directly determined. It is expected that the attitude of the Cluster will be known to better than $0.5°$ using these measurements. The exact performance to be expected in all mission segments will be studied in Phase A.

The Deep Space Network (DSN) of the Jet Propulsion Laboratory (JPL) will provide tracking, telemetry and command (TT&C) support for the ISC Mission.

Data transport services will be performed by the NASA Communications (NASCOM). Other ground facilities needed to support ISC include Payload Operations Control Center (POCC), Command Management System (CMS), Flight Dynamics Facility (FDF), Network Control Center (NCC), and the Data Capture Facility (DCF). The science ground system will be equipped with a Central Data Handling Facility (CDHF) for generation of higher level data product and a Science Operation (SOC). The SOC will be used as the interface with the Principal Investigators (PI's) and Co-Investigators (Co-I's) for science planning and operations. The PI and other science users will receive science data products and ancillary data from the CDHF for further data processing and analysis at the remote sites.

Commands initiated by the ISC operational staff will be uplinked to the spacecraft via the POCC and the DSN.

Telemetry data received by the DSN will be transmitted to the POCC for spacecraft health and safety processing, on the FDF for spacecraft orbit/attitude and maneuver processing support and to the DCF for science data capture and level zero processing. The POCC is the focal point for mission providing mission planning/scheduling, spacecraft health and safety monitoring. The CMS receives observation request by the PI's and Co-I's, analyzes the request against spacecraft constraints, and generates the command loads for delivery to the POCC for uplinking.

Spacecraft tracking and telemetry data are received by the FDF through NASCOM. Science data is transmitted through NASCOM to the DCF, where data is captured and level zero processed. The DCF also provides data accounting and data storage services, and data distribution to the CDHF. The CDHF performs high-level data processing on the received science data. The PI's, Co-I's, and other science users will receive science data products and ancillary data for further processing and analysis, on a request basis.

Technology Requirements:

Inter-spacecraft relative positions to 1% and attitude to 0.01° for cluster spacecraft; inter-spacecraft communication; imagers using O+ resonantly scattered radiation; X-ray imager improvements; improved imaging of He+ resonantly scattered radiation at 30.4 nm; continued development of ENA cameras.

Points of Contact:

Science Panel:	Dr. Daniel N. Baker (301) 286-8112
	Code 690
	Goddard Space Flight Center
	National Aeronautics and Space Administration
	Greenbelt, MD 20771
Discipline:	Magnetospheric physics
Program Manager:	TBD
Program Scientist:	Dr. Tom Armstrong (202) 453-1514
NASA Center:	GSFC
Project Manager:	Steve Paddack (301) 344-4879
Project Scientist:	TBD

Section 6 Candidate Future Missions

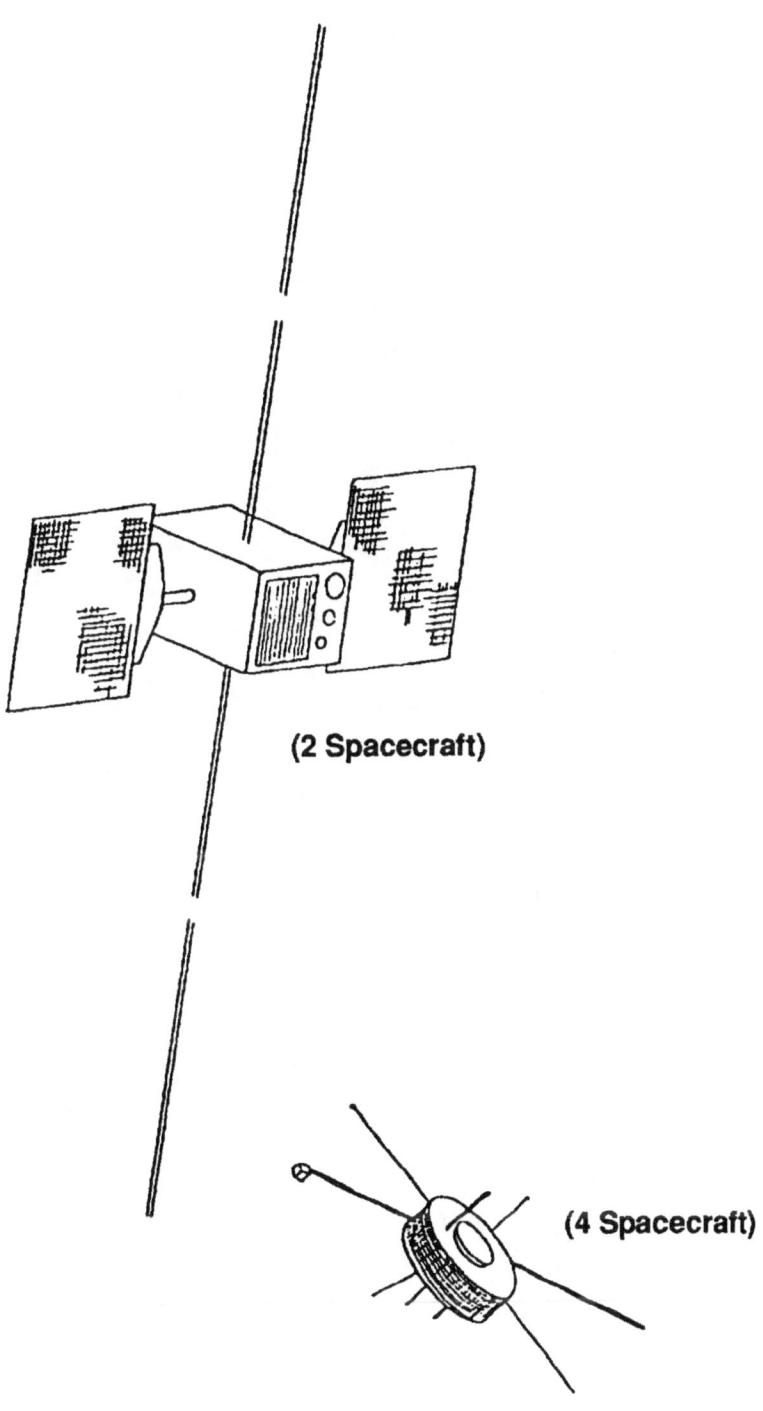

(2 Spacecraft)

(4 Spacecraft)

Imaging Super Cluster

6.10 Grand Tour Cluster (Magnetopause/Near-Plasma Sheet Cluster)

Target: Magnetopause/magnetotail

Orbit: 0° inclination 2 x 12 R_e
0° inclination 2 x 30 R_e
23° inclination 8 x 80–235 R_e double lunar swingby
90° inclination 2–10 x 30 R_e

Mission Duration: ~6 years

Mission Class: Moderate

Mass: 400–500 kg each

Launch Vehicle: Atlas II-A or Atlas II-AS

Theme: Magnetospheric physics

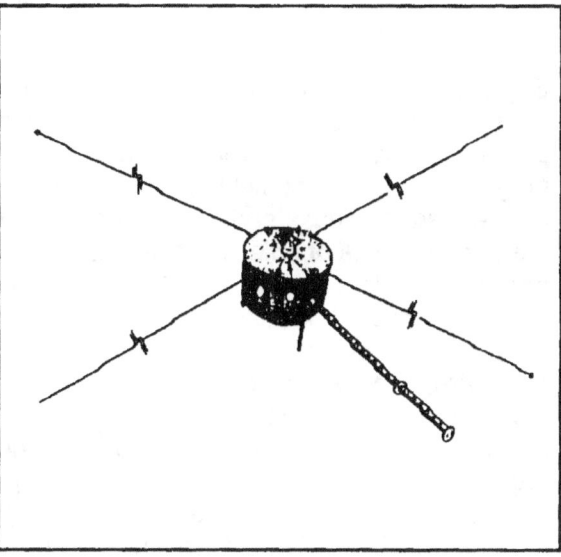

Science Objectives:

Just as the ISTP program will provide the first global study of the magnetosphere, the GTC will provide the first comprehensive study of the micro- and mesoscale processes of the magnetosphere. This represents a natural extension of the ISTP mission. We emphasize, however, that the physics underlying the GTC represents an entirely new frontier when compared to ISTP. Specifically, the physics to be studied in the ISTP will center on the global transfer of energy in the magnetosphere. Once this overview has been obtained, the logical next question concerns the nature of the physical processes that underlie the transfer of energy between various regions of the magnetosphere. ISTP will offer some preliminary insights into these questions, but is simply not equipped to answer questions that require information on small and intermediate length scales. A different type of mission is required and that is the GTC.

It is envisioned that near the end of the ISTP mission the interaction between theory and experiment will have produced the next generation of quantitative pictures of the global physics of the magnetosphere. The next questions to be asked will involve an investigation of the mesoscale physics that characterize magnetospheric boundary layers and critical regions of the magnetosphere.

Much as the development of global three dimensional magnetohydrodynamic (MHD) simulation codes in the decade preceding ISTP provided an impetus to study global magnetospheric physics in a comprehensive manner, the development of three dimensional hybrid codes which allow scientists to resolve the physics of mesoscale magnetospheric processes during the early to mid 1990's will set the stage for the next generation of magnetospheric missions. These missions will require the use of clustered spacecraft to study widely separated magnetospheric regions. The GTC would offer a moderate cost, first step in this new era of magnetospheric exploration. The GTC would provide the data for learning about the global interaction of mesoscale processes. This could be coupled with the miniaturization expertise to be gained in the proposed Auroral Cluster Mission to design a highly cost effective simultaneous global survey of the magnetosphere using multiple clusters in the next century.

A detailed discussion of the scientific objectives of the individual segments of the GTC mission follows.

Segment 1. The magnetopause is the site of mass, energy, and momentum transfer between the solar wind and the magnetosphere. We are currently faced with a bewildering array of possible steady- and transient-interaction mechanisms. Each predicts a specific pattern of magnetic-field, electric-field, plasma, and energetic particle signatures as functions of position on the magnetopause surface. The sub-solar or equatorial magnetopause is believed to be the dominant site of solar wind-magnetosphere interaction: prolonged observations are essential in this region. To date there has been no comprehensive survey of processes occurring at very low latitude, and the contemporary data cannot uniquely distinguish among the possible processes or their relative importance. The inner edge of the plasma sheet and the near-magnetotail is the region where the substorm process may be initiated. In this region, the magnetic field makes the transition from dipole-like to tail-like, and substorm plasma injection occurs. Although the region near geosynchronous orbit has been well explored, the region from ~8–12 R_e is very much underexplored. This area must be surveyed in order to understand the processes that produced particle injections, the disruption of the cross-tail current, the change in field topology and the mapping of field-aligned currents.

In Segment 1, the GTC will provide understanding of the important physical processes operating both at the magnetopause and in the near tail; a cluster of at least four properly instrumented spacecraft is required. Utilizing variable separations between spacecraft, the cluster will make it possible to separate spatial from temporal variations and to determine the spatial morphology of the structures of interests. This mission will investigate the coupling of magnetospheric domains and, in particular, will focus on domain boundaries and coupling across spatial and temporal regimes.

Some specific objectives of Segment 1 are:

- Study in detail magnetopause structure and processes (FTEs, pressure pulses, surface waves, impulsive entry, steady-state or enhanced merging).

- Examine solar wind and IMF control of each of these processes.

- Resolve space-time ambiguities by lingering in the equatorial magnetopause region with a high time-resolution, variably-spaced cluster.

- Perform a detailed investigation at all local times on the day-side.

- Examine the near-tail region with cuts through the transition region from dipolar to tail-like geometry.

- Determine the processes involved in sub-storm particle injection.

- Study the process of current sheet interruption.

To address these science objectives, measurements of the following are needed: DC and AC electric and magnetic fields, 3-D ion composition and electrons (few eV to 50 keV), cold electron density and temperature and density fluctuations, and energetic-ion composition (up to few MeV/nucleon) and electrons (to 1 MeV). An upstream monitor of the solar wind is required.

Segments 2 and 3. These segments consist of the GTC super in an elliptical magnetotail orbit with apogee that ranges from ~30 R_e to 200 R_e with variable inclination. This mission provides:

- The three-dimensional topology of the magnetosphere, and the plasmasphere, their boundaries and their dynamics on a local to mesoscale.

- A measure of magnetotail boundaries, current systems, and their relation to inner magnetosphere/ionosphere boundaries and current system on a local scale to mesoscale.

The cluster will provide *in situ* magnetotail observations that allow 3-D determination of boundaries and current systems throughout the magnetotail. The GTC will separate spatial and temporal effects; by using a variable satellite spacing (from a fraction of an R_e to ~10 R_e) and will be capable of observations out of the plane of the nominal cluster orbit.

Segments 2 and 3 of the GTC will represent a major advance in our goal of understanding the behavior of the role of local processes in determining the observed global features. It builds upon and is a natural extension of the ISTP program, which is studying the mass, momentum, and energy flow through the magnetospheric system. An understanding of this flow will provide a basis on which to interpret and test the ISTP results concerning global dynamics and their relation to local processes.

Segment 4. During this phase of the mission the GTC is intended to explore in depth three important boundary regions of the magnetosphere: the day side magnetopause on the night side and the polar cusp; and the region near 10 R_e in the tail, which is the transmission region between the near-Earth ring current and the tail plasma sheet. The orbit would initially be R_e circular polar orbit, with a local time of the orbit plane at about 1500–0300 hours. At this distance and local time, the cluster would spend about 15 hours skimming the day side magnetopause from the southern to northern cusp every two days, crossing over the pole and then traversing the tail from north to south and returning to the day side. The orbit, virtually inertially stable, would decrease in local time by two hours per month, reaching 0900–2100 Magnetic Local Time (MLT) after three months. At this point, the day side apogee could be gradually increased to match the flare of the dawn-side magnetopause, becoming a 10 x 30 R_e orbit when it reaches the dawn-dusk meridian. At this time the dawn-side of the orbit would be skimming the magnetopause while the dusk-side portion would remain in the low-altitude boundary layer, investigating particle-entry mechanisms. Propulsion could also be used to tilt the line of apsides up out of the ecliptic plane, reaching perhaps 35° and an apogee of 30 R_e after 6 months in Segment 4. Now the day-side (perigee) portion would spend about a day skimming the day-side magnetopause from pole to pole (being slightly inside the average magnetopause location in the southern hemisphere and outside of it in the northern hemisphere), and the night-side portion would cross the magnetopause considerably down the tail, once every 5.6 days. In this way, momentum would transfer (and supposed northward IMF reconnection) could be investigated in a near-skimming trajectory. The cluster would spend

Section 6 Candidate Future Missions

about three months traversing the tail in this manner, and then would swing back around to the day-side, investigating the high-altitude bow shock as well.

This segment would be highly synergistic with the earlier Segment 1 of the mission, which was flown in the near-Earth equatorial plane (2 x 12 R_e orbit). In contrast, the spacecraft in Segment 4 will cut through the magnetotail from north to south. Whenever the GTC was out in the solar wind at apogee (about half the time for half of the year), it could serve as high-quality solar wind monitor for other magnetospheric missions, much closer to the magnetopause than, say, a Wind orbit. It will also serve as an excellent solar-wind turbulence investigation.

The GTC Segment 4 breaks new territory by being the first mission to skim the day-side magnetopause and the turbulent exterior cusp. It will be the first to cross the polar magnetopause at distances tailward of $X \sim -10\ R_e$, crossing it at 20 R_e or more. In this way it will be the first to explore the magnetopause cross-section at high latitudes behind the Earth.

Day-side magnetopause skimming orbits will make it possible to follow the development of plasma which is injected either by FTE's, spatially and/or temporarily varying day-side merging, or quasi-diffusive processes.

In the near-tail, the mission can monitor the fate of the plasma mantle: does the plasma mantle eventually reach the tail neutral sheet and return as the plasma sheet, or does its supersonic flow never return? In addition, the mission can trace the fate of ionospheric plasma fountains emitted from the day-side cusp and night-side auroral zone, testing where they convect back to the neutral sheet and providing quantitative tests of the relative importance of mantle plasma versus low-latitude boundary layer and ionospheric plasma in supplying the plasma sheet.

For substorms processes, Segment 4 will provide an out-of-ecliptic monitor to check the size and motion of plasmoids. Segment 4 will also, for the first time, allow measurements of the vertical dimension and field strength in the lobes to measure total magnetic flux changes during growth phases and expansion phases of substorms.

The polar magnetosheath will also be skimmed for part of the mission. At this part of the trajectory, one can monitor solar-wind deceleration by current dynamo processes, and perhaps watch bursts of magnetospheric plasma, which will be most easily observed in the magnetosheath just outside the magnetopause. The GTC will spend about 1 day in its 5.6-day orbit in that region, being engulfed periodically by the magnetopause as it "breathes" and/or flaps.

Spacecraft:

    Type: Spinners (5 S/C)

    Special Features: Inter-spacecraft ranging

    Special Requirements: Relative position accurate to 1% and attitude to 0.01°; continuous solar wind monitoring required; auroral & plasma imaging desired (from Inner Magnetosphere Imager)

For purposes of redundancy, lifetime and guaranteed three dimensional coverage, the GTC is proposed as five spacecraft.

The GTC satellites will be identical and stabilized spin with ISTIP, ISEE and DE heritage. It is estimated each spacecraft will be an approximately 1/2 scale Wind spacecraft of the ISTP configuration with a reduced instrument complement. Approximately 40% of the launch mass of each spacecraft is fuel.

Instruments:

	Mass	Power	Data rate	Data storage	FOV
Electric Fields	TBD	TBD	TBD	TBD	TBD
Magnetic Fields	TBD	TBD	TBD	TBD	TBD
3-D Electron & Ion Distributions with Composition	TBD	TBD	TBD	TBD	TBD
Energetic Particles & Thermal Plasma with Composition (1 S/C only)	TBD	TBD	TBD	TBD	TBD

High bit-rates (up to MHz) required for short periods of time in the vicinity of the magnetopause. Could be accomplished with large on-board data storage and slower playback.

For the baseline payload we assume the same instrument complement as on the ISTP Cluster. Each of the spacecraft will be identical set of instruments and will provide Measurements of Magnetic fields (DC to 10 kHz), Electric fields (DC to 1 MHz), 3-D electron and ion distribution with composition (few eV to 30 keV) and energetic particles and thermal plasma and composition.

Mission Strategy:

A Grand Tour Cluster Mission (GTC) is proposed. It would be composed of a minimum of four identical spacecraft to be flown in a tetrahedral configuration over at least part of each segment of the proposed mission. For the purposes of redundancy and guaranteed three dimension coverage, 5 spacecraft are preferred. The prime mission objective would be to distinguish between temporal and spatial variations in critical regions and boundary layers of the terrestrial magnetosphere for the first time. The movement of the cluster of spacecraft between widely separated parts of the magnetosphere would be accomplished with a combination of lunar assists and on-board propulsion. The sequence of mission segments has been chosen so as to minimize the on-board propulsion requirements. The proposed mission would represent a major advance over the present ISTP Cluster mission in a number of respects: (1) the inter-satellite positions would be determined with much greater accuracy, (2) a far more complete survey of the magnetosphere would be accomplished, (3) a much greater range of spatial scales would be sampled. The mission would be composed of four segments each of

Section 6 Candidate Future Missions

duration of approximately two years. These segments in their order of execution are: (1) an equatorial magnetopause, inner magnetotail survey, (2) a mid-magnetotail survey, (3) a deep magnetotail study, (4) a study of the higher latitude magnetopause. Much as the ISTP Cluster will provide a scientific and engineering heritage for the GTC, the GTC will lay the foundation for future missions employing multiple clusters to simultaneously study the magnetosphere. The GTC would perform the first complete microscale and mesoscale reconnaissance of the magnetosphere.

The GTC could either be precursor mission to the Imaging Super Cluster (ISC) spacecraft, or depending on the time scales and budgetary constraints, the GTC could be a component of the ISC. We note that the moderate cost of the GTC compared to the major-mission status of the ISC, as well as the advantages of simultaneous multi-clusters, make persuasive arguments for the GTC as a precursor of the pathfinder mission. This would allow an initial survey of the important boundaries and regions of the terrestrial magnetosphere while the technologies for the imagers and those for the miniaturization for cost-effective multi-clusters are developed. It would also better pose the questions to be answered by the ISC and hence will have beneficial effects on the ISC mission design.

For the purpose of comparison with the Future Magnetosphere Missions proposed by the Magnetospheric Physics Panel of the Space Physics Strategy Implementation Study, a brief description of the relationships follows.

The first segment duplicates the proposed Magnetopause/Near-Magnetotail Equatorial Cluster Mission. The second and third segments cover the cluster measurements to be performed as part of the Imaging Super Cluster. The fourth segment would cover the Magnetopause/ Boundary Layer Explorer mission.

The sequence of orbital transfer is given below. Each segment is to be executed for a minimum of two years. The tetrahedral configuration should be maintained for at least part of each segment subject to fuel consumption economies. It is proposed that for the initial orbits of each segment, the tetrahedral configuration be held throughout. After an initial survey, the priorities for holding this configuration during the segment can be established. Typical separation distances for each segment is also shown.

Mission segment	Perigee x apogee	Separation
1—Low equatorial, 0°inclination	2 x 12 R_e	1 km–0.1 R_e
2—High equatorial, 0° inclination	2 x 30 R_e	0.01–1.0 R_e
3—Deep tail (double lunar swingby)	8 x 80–235 R_e	1–10 R_e
4—Polar, 90°	10 x 30 R_e	1 km–1.0 R_e

In the above table, position accuracy as a function of geocentric distance is given. A goal is to have a total relative position accuracy among the cluster satellites known to about 1% or 100 m whichever is larger.

Using propulsive maneuvers, the five satellites will be actively positioned relative to each other so as to measure the time varying, three-dimensional magnetosphere at the GTC. The requirement is to know this separation distance to an accuracy of 1% or 100 meters, whichever is larger. An inter-satellite tracking system will be included on each satellite to determine these relative positions for separations less than about 1 R_e. Normal ground-based tracking will do the determination when the separations are larger.

The selection of an inter-satellite tracking system is based on the fact that at the shortest separations, this measurement is virtually impossible to perform using ground-based tracking techniques because of the poor geometric relationships from the satellites and the ground station. Also, it is desirable that these measurements be made on a continuous basis. For some of the lower-altitude segments, limited viewing opportunities from the three DSN stations make this impossible. Even when stations can see the GTC, it is not feasible to dedicate so much time on the DSN to this function. These reasons lead to the use of a radio frequency tracking technique carried by the satellites themselves.

Another requirement is that when any one of the satellites detects the beginning of a scientifically interesting event, it must send a trigger signal to the other satellites of the cluster. The trigger signal changes the clusters' mode of operation. The time between detection of the event by one satellite and triggering all the other satellites to the desired mode must be less than one second and, as a goal, less than 0.1 second. The inter-satellite separations will also serve to send this trigger signal.

To minimize the amount of extra equipment each satellite carries to perform the inter-satellite tracking function, maximum use will be made of the S-band Deep Space Net (DSN) tracking transponder that is required for communications and tracking. The raw tracking information will be collected by the satellites and the calculations of the inter-satellite separations will made on the ground.

One implementation approach is as follows: Each satellite will be equipped with a low power S-band transmitter that is set to the same frequency as the S-band uplink from the DSN. The Cluster is programmed so that each satellite in its turn broadcasts a 3 Mbps pseudo-noise (PN) code sequence over this link. Each satellite uses the same PN code, but with a different phase shift so that all the PN codes are uncorrelated with each other. A low rate data channel for satellite identification and the trigger signal is also sent.

The four satellites that are not transmitting lock up on the satellite that is transmitting and each measures the relative time (phase) between its own internally generated PN code and the received PN code. Since each satellite knows when to expect a given satellite to broadcast and what the relative phase measurements were for the last few contact, it is able to rapidly lock on to the received PN code. It is expected that a transmission of less than one second's duration, repeated once a minute by each satellite in the cluster, will be sufficient when the cluster is closely spaced. Longer transmissions may be required at the longest separations. When a satellite is performing a propulsive maneuver, it may be scheduled to increase the duration and/or the rate of its transmissions as necessary.

If a satellite determines that a scientifically interesting event has occurred, it simply turns on its inter-satellite transmitter and sends the "trigger even" message as a low data-rate word along with the normal identification message. (If already transmitting, changes the word's contents to indicate a trigger function has occurred.) The exact signal structure for this trigger function is to be determined, but it must be secure and not subject to false triggering. In other words, it should incorporate the same safeguards as a command.

When a satellite is not transmitting a tracking message, and when not tracking a scheduled transmission from another satellite, each receiver looks for a trigger signal from any of the satellites in the cluster or a command transmission from the ground. When a trigger message is received and verified to be correct, the satellite executes the preprogrammed mode change.

When a signal from the ground is received, each satellite determines whether or not the signal is addressed to itself. If yes, it would not transmit its inter-satellite tracking signal as scheduled, but would execute whatever functions the command message told it to do. If the message is for one of the other satellites, it ignores the command message and continues with the inter-satellite tracking transmissions on the usual schedule.

Whenever a satellite is neither transmitting nor receiving tracking messages, it alternately looks for a trigger signal from another satellite and a ground transmission addressed to itself.

Since the ground transmission also uses a PN code, each satellite detecting the signal notes the phase difference between the received PN code and its own PN code. These measurements are made no matter what satellite the message is addressed to. The ground command message and the inter-satellite tracking signals would not interfere with each other because of the different PN code correlations between them.

Each satellite's telemetry link to the ground normally uses its own PN code. The ground station notes the time difference between this PN code and its own PN code. The satellite is also commanded to retransmit or echo the uplink PN code. The time difference measured on the ground then represents a two-way range measurement.

Note that the DSN does not now support PN-coded telemetry or command systems, but they have expressed interest in using these techniques in the future. There is also the possibility that S-band will be phased out because of terrestrial interference. The PN modulated system may be able to combat this interference and so extend the life of S-band communications. If S-band is not available when the GTC is to fly, the other DSN frequencies will be used.

Since the tracking technique described is similar to the 30 to 100 meter accuracy demonstrated by the C-code used in the Global Positioning System (GPS), it is expected that the inter-satellite accuracy will also achieve 100 meter or better accuracy.

Because of the overdetermination of the inter-satellite separations and the ranges to the ground station, one can solve for the various oscillator drifts. It is not necessary for the GTC to use the atomic clocks flown on the GPS satellites. Only reasonably good crystal oscillators are required.

Since the PN codes from the GTC are uncorrelated with each other, it is possible to receive telemetry data simultaneously from all the GTC satellites that are within the receiving antenna beam. For many spacings and distances, all five satellites can be serviced at one time.

It is also important to know the orientation of the Cluster relative to inertial space. This will be determined using all the inter-satellite and ground-tracking data and orbit mechanics. A given ground measurement will directly measure two angles of the Cluster's orientation relative to the line of sight between the station and the Cluster. It will not directly measure the third angle which is the orientation of the Cluster "around" the line of sight. When the relative orientation between the Cluster and the station changes by a large angle, then the missing orientation angle may be directly determined. It is expected that the attitude of the Cluster will be known to better than 0.5° using these measurements. The exact performance to be expected in all mission Segments will be studied in Phase A.

The Deep Space Network (DSN) of the Jet Propulsion Laboratory (JPL) will provide tracking, telemetry, and command (TT&C) support for the GTC mission.

Data transport services will be performed by NASA Communications (NASCOM). Other ground facilities needed to support GTC include a Payload Operations Control Center (POCC), a Command Management System (CMS), a Flight Dynamic Facility (FDF), a Network Control Center (NCC), and a Data Capture Facility (DCF). The science ground system will be equipped with a Central Data Handling Facility (CDHF) for generation of higher-level data products and a Science Operation Center (SOC). The SOC will be used as the interface with the Principal Investigators (PI's) and Co-Investigators (Co-I's) for science planning and operations. The PI and other science users will receive science data products and ancillary data from the CDHF for further data processing and analysis at the remote sites.

Commands initiated by the GTC operational staff will be uplinked to the spacecraft via the POCC and the DSN.

Telemetry data received by the DSN will be transmitted to the POCC for spacecraft health and safety processing, to the FDF for spacecraft orbit/attitude and maneuver processing support and to the DCF for science data capture and Level zero processing. The POCC is the focal point for mission operation providing mission planning/scheduling, spacecraft health and safety monitoring. The CMS receives observation requests by the PI's and Co-I's, analyzes the requests against spacecraft constraints, and generates command loads for delivery to the POCC for uplinking.

Spacecraft tracking and telemetry data are received by the FDF through NASCOM. Science data is transmitted through NASCOM to the DCF, where data is captured and Level Zero processed. The DCF also provides data-accounting and data-storage services, and data distribution to the CDHF. The CDHF performs high-level data processing on the received science data. The PI's, Co-I's, and other science users will receive science data products and ancillary data for further processing and analysis, on a request basis.

Technology Requirements:

Inter-spacecraft relative position accurate to 1% and altitude to .01% and altitude to .01°; inter-S/ C links.

Section 6 Candidate Future Missions

Points of Contact:

Science Panel:	Dr. Daniel N. Baker (301) 286-8112
	Code 690
	Goddard Space Flight Center
	National Aeronautics and Space Administration
	Greenbelt, MD 20771
Discipline:	Magnetospheric physics
Program Manager:	TBD
Program Scientist:	Dr. Tom Armstrong (202) 453-1514
NASA Center:	GSFC
Project Manager:	Steve Paddack (301) 344-4879
Project Scientist:	TBD

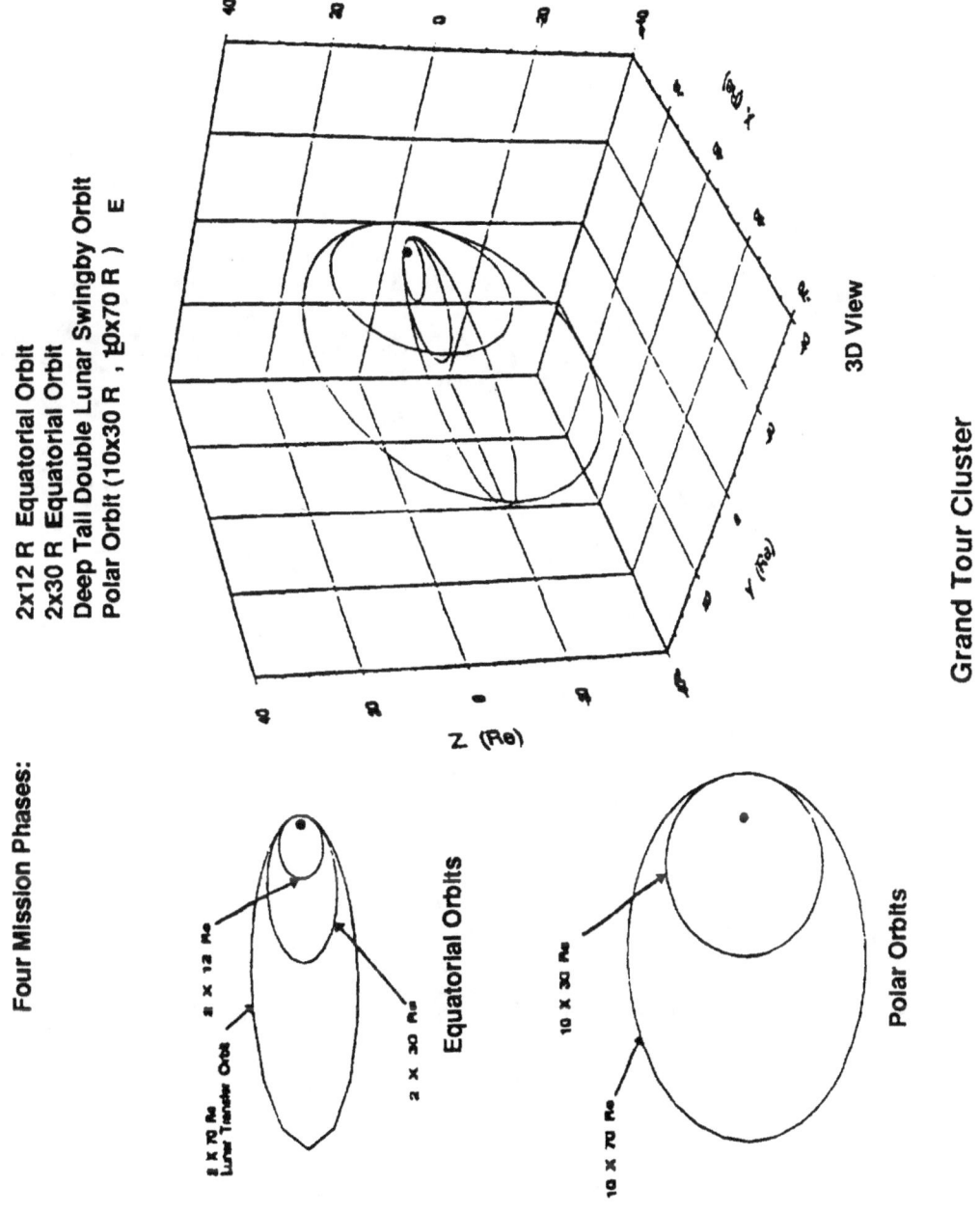

Grand Tour Cluster

6.11 High-Energy Solar Physics (HESP)

Target: The Sun and selected non-solar objects

Orbit: 600 km
Low inclination

Mission Duration: 3 years

Mission Class: Intermediate (Explorer)

Mass: 3450 kg

Launch Vehicle: Medium ELV

Theme: Solar high-energy astrophysics/solar flares

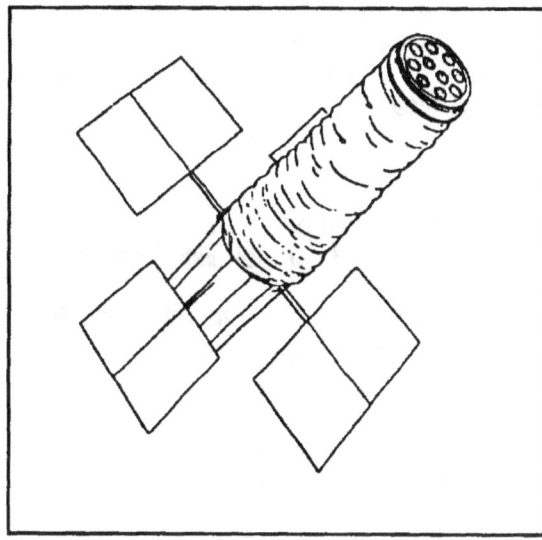

Science Objectives:

The ability to release energy impulsively through particle acceleration to high energies is a common characteristic of cosmic plasmas at many sites throughout the universe, ranging from planetary and stellar magnetospheres to active galaxies. These high-energy processes play a central role in the overall physics of the system at each site where they are observed. A detailed understanding of these processes is one of the major goals of astrophysics, but in essentially all cases, we are only just beginning to perceive the relevant basic physics.

Nowhere can one pursue the study of this basic physics better than in the active Sun, where solar flares are the direct result of impulsive energy release and particle acceleration. The acceleration of electrons is revealed by hard X-ray and gamma-ray *bremsstrahlung;* the acceleration of protons and nuclei is revealed by nuclear gamma-rays and neutrons. The accelerated particles, notably the electrons with energies of tens of keV, probably contain a major fraction of all the released flare energy.

The primary scientific goal of HESP is to understand the impulsive release of energy, efficient acceleration of particles to high energies, and rapid transport of energy. These fundamental processes occur in solar flares, supernova remnants, the galactic center region, active galaxies, and other sites throughout the universe. Solar flare studies are the centerpiece of our investigation because in flares these high energy processes not only occur routinely but also can be observed in unmatched detail in hard X-rays and gamma-rays. With the powerful new instrument on HESP, we expect to locate the regions of particle acceleration and energy release, characterize in great detail the accelerated particle distributions as functions of both space and time, follow the subsequent transport of energy through the plasma, and thereby identify the operative physical mechanisms.

A large solar flare releases as much as 10^{32} ergs in times as short as 100 to 1000 seconds. It is believed that the energy comes from the dissipation of the non-potential components of strong magnetic fields in the solar atmosphere, possibly through magnetic reconnection. Of the many available signatures of this energy release, hard X rays, gamma-rays, and neutrons form a distinct class in that they are produced before the accelerated particles are thermalized in the ambient atmosphere. Consequently, they provide the most direct information available on the energy release and particle acceleration processes.

Solar Flare Studies

- Identify the particle acceleration mechanisms at work during different phases of flares and coronal disturbances.

- Determine the contributions of high-energy particles to flare energetics, specifically by following the *bremsstrahlung* and nuclear line radiations spatially and temporally.

- Study particle transport (distribution functions, drifts, and scattering) during flares and coronal disturbances.

- Study abundances and abundance variations in the solar plasma as revealed by nuclear line intensities.

Galactic and Extragalactic Gamma-ray Source Studies

- Localize the hard X-ray and gamma-ray sources in the galactic center region, including the positron-annihilation source and repeating soft gamma-ray bursters.

- Determine the energetic particle distributions in supernova remnants including the Crab.

- Study the spatial structure of hard X-ray emission in clusters of galaxies.

- Monitor line emission and other features in the spectra of sources such as the SS433, the Crab pulsar, and the galactic center.

- Study the spatial structure of active galactic nuclei at energies above 10 keV, including determination of the ratio of core to jet emission.

Spacecraft:

    Type: Spinner

    Special Features: TBD

    Special Requirements: Fine guided with <10 Mbps telemetry system

Instruments:

	Mass (kg)	Power	Data rate	Data storage	FOV
High Energy Imaging Spectrometer	1550	TBD	TBD	TBD	Whole Sun

The HESP payload consists of the High-Energy Imaging SPECtrometer (HEISPEC). HEISPEC will be capable of obtaining hard X-ray and gamma-ray images of solar and celestial sources with arcsecond angular resolution and keV energy resolution from 10 keV to 20 MeV. It will also provide imaging spectroscopy of solar flare neutrons from 20 MeV to 1 GeV and moderate resolution gamma-ray spectroscopy of energies in excess of 100 MeV. It has high sensitivity and sub-second time resolution for solar flare observations and it will be capable of imaging many non-solar sources of hard X-ray emission. For a deep (non-solar) observation lasting 10^5 s, it will achieve a 3-sigma sensitivity for each subcollimator of 4×10^{-6} photons cm^{-2} s^{-1} keV^{-1} at 100 keV.

To date, solar hard X-ray images have been limited to a resolution of 8 arcseconds and to energies below 40 keV. HEISPEC will have 10 times better angular resolution, the capability of imaging solar flares to energies as high as 10 MeV, and a factor of 1000 improvement in sensitivity over the Hard X-ray Imaging Spectrometer (HXIS) on SMM. The high sensitivity will, for the first time, allow studies of solar flares to be made on arc second size scales and subsecond time scales, comparable to those associated with the magnetic structures and with the processes that modify the high energy electron spectrum. The high sensitivity and full Sun coverage of HEISPEC will allow the spatial study of many flares each day, thus providing a large sample of high-energy flares for statistical analysis.

In addition to its fine angular resolution, HEISPEC will have spectral resolution sufficient for accurate measurement of all parameters of the expected solar gamma-ray lines with the exception of the neutron capture deuterium line, which has an expected FWHM of about 0.1 eV. This capability is illustrated a figure at the end of this section where the spectral resolution of the HEISPEC 2-segment High Purity germanium (HPGe) detectors is compared with the typical line widths expected for gamma-rays in solar flares. Also shown for comparison is the resolution of the hard X-ray and gamma-ray spectrometers on SMM and the resolution required to resolve the steep super-hot thermal component of solar flares.

HEISPEC is based on a Fourier-transform imaging technique. Two widely spaced, fine-scale grids temporally modulate the photon signal from a source close to the axis as the individual collimators or the whole system rotates about that axis. The modulation can be measured with a detector having no spatial resolution placed behind the second grid. This is a classical rotating modulation collimator (RMC) design of the type used on previous missions such as the US–SAS–C and the Japanese Hinotori spacecraft. The modulation pattern contains information on the amplitude and phase of many spatial Fourier components of the image for the finite range of source sizes. Multiple RMC's, each with different slit widths, provide coverage of different ranges of source sizes. An image is constructed from the measured Fourier components in exact mathematical analogy to multi-baseline radio interferometry. The use of HPGe detectors to measure the modulation allows us to determine the spectrum with high energy resolution at the same time that the image is being determined with fine spatial resolution.

The table below summarizes the HEISPEC instrument parameters. The basic design uses 10 grid pairs with different slit widths to cover source size scales from the arc second range to several arc minutes. Each grid pair sub-collimator has three HPGe detectors to determine the energy of each photon that passes through the RMC. These detectors are n-type coaxial HPGe detectors cooled to 90 K and surrounded by bismuth germanate (BGO) and plastic scintillator anticoincidence shields to minimize the background. Electrical segmentation of each HPGe detector into a thin front segment and a thick rear segment, together with a pulse-shape discrimination, provides optimal dynamic range and signal-to-background characteristics for flare measurements over the energy range from 10 keV to 20 MeV. Neutrons and $\geq$ 20 MeV gamma-rays are detected and identified with the combination of the HPGe detectors and the rear BGO shield.

HEISPEC Instrument Parameters

Imaging	
Technique:	Fourier-transform imaging using multiple RMCs
Angular resolution:	$\leq$ 1 arcsecond
Field of view:	Full Sun (0.6°)
Boom length:	$\geq$ 10 m
Parameters of grids:	
No. of grid pairs:	10
Slit spacings:	$\leq$50 microns to 6 mm (1 arcsecond to 3 arcminutes)
Thickness:	$\leq$ 0.5 to 4 cm (depending on slit spacing)
Material:	Tungsten or other high Z material
Spectroscopy	
Energy range:	
Hard X-rays & Gamma-rays:	10 keV 20 20 MeV (HPGe)
	20 MeV to 100 MeV (BGO)
Neutrons:	20 MeV to 1 GeV
Energy resolution:	0.6 to 5 keV FWHM (HPGe)
Total detector area	1000 cm^2
Shield	5-cm thick BGO on sides and at rear
HPGe cooling:	Stirling Cycle cooler to 90K

Mission Strategy:

Launch in 1998 before rise in activity to maximum in cycle 23. If possible, simultaneous operation with SOHO and OSL for optimal definition.

The spacecraft will be placed in a low-inclination orbit at 600 km by a MELV launched from the Eastern Space and Missile Center. On orbit the telescope will deploy to allow for a 10-meter separation between the upper and lower grids. Upper-to-lower grid registration will be monitored by sensors attached to grid structure support and controlled by mechanisms on the upper and/or lower grid frames. The S/C will be Sun-pointed and spun at 15 rpm. A counter-rotating momentum wheel will balance the angular momentum produced by the spinning spacecraft. The transverse moment of inertia produced by the deployed solar arrays and balance weights will be such that the same moment of inertia exists about all axes so as to minimize gravity gradient torques. The satellite could be built without a counter-rotating momentum wheel; however, for stability the transverse inertia would have to be greater resulting in larger balance weights and/or long booms.

Causing a spacecraft to spin is one method of producing the modulation for the Fourier imaging. Other methods are conceivable; *viz.*, a spinning instrument on a despun spacecraft (three-axis stabilized) or separately rotating and synchronized grid pairs on a three-axis stabilized instrument/spacecraft. All of these other methods should be investigated in depth during the Phase A Study.

Fine pointing is not required; the optical axis of the telescope must be pointed at the target to within 3 arc minutes. Similarly, high pointing stability is not required provided that the knowledge of the orientation of the optical axis is available to ~0.1 arcsecond and millisecond time scales. This knowledge can be obtained for solar-pointed observations from measurements of the solar limb and for non-solar observations from the spacecraft fine error system.

All normal operations are to be supported through the TDRSS. The designated ground stations will be used for contingency operations when the TDRSS is not available. The packetized telemetry will be transmitted to the Goddard Space Flight Center (GSFC) ground operations via TDRSS or the designated ground stations. Data transport services will be provided by the NASA Communication (NAASCOM). Other ground facilities needed to support the HESP include a Payload Operations Control Center (POCC), a Command Management Systems (CMS), a Flight Dynamics Facility (FDF), a Network Control Center (NCC), and a Data Capture Facility (DCF). The PI's and CO-I's will receive raw science data or data products and necessary ancillary data required for further data processing and analysis at their home facilities, and these data together with the necessary software will be archived at the Nation Space Station Data Center (NSSDC). Other science users can access the HSDPC for science data archives through a data catalog and browse system one year after generation of data products by the PI's and Co-I's.

HESP commands initiated from the operational staff will be uplinked via the POCC at 1 kbps via Tracking and Data Relay Satellite System (TDRSS) Multiple Access (MA).

The HESP will contain its own Communications and Data Handling (C&DH) system. This spacecraft subsystem will store pre-planned command sequences for HESP operations. Engineering data from the spacecraft is estimated to be 10% of science data. The packetized science and engineering data will be transmitted once per orbit when the spacecraft is in TDRSS view. The on-board data storage system will be able to store up to 2 orbits of data.

Ground data operations and processing elements for the HESP include the White Sands Ground Terminal, and the mission operation and data processing facilities mentioned above. The Investigator team will assist with the ground mission operations and data processing activities. HESP data transport support is provided by NASCOM which interacts with the POCC and NCC. The NCC enables HESP users to TDRSS services and ensures the quality of services.

Telemetry data is dumped as required to the POCC for spacecraft housekeeping data, FDF for spacecraft orbit/attitude data, and to the DCF for science data. The POCC is the mission control center, providing mission planning/scheduling, spacecraft health and safety monitoring, end-to-end testing, simulation, and analyzing customer requests such as checking with network control and TDRSS support. The CMS receives observation requests for the spacecraft by the PI and Co-I's, analyzes requests against spacecraft constraints, and generates the command sequence for approved requests for delivery to POCC for uplinking.

Spacecraft orbit and attitude is received by the FDF through NASCOM. The POCC and CMS interact so that spacecraft commands may be uplinked to the HESP. Packetized science data is transmitted through NASCOM to the DCF, where data is captured and Level Zero processed. The DCF also provides data accounting and data storage, and data distribution to the HSDPC.

Technology Requirements:

During the Phase A study a number of subjects will be addressed:

- Spin modulation/spacecraft options will be examined in depth to arrive at a preferred method.

- The application of Controlled Structure Interaction (CSI) technology to the deployment and control of the grid alignments will be assessed.

- Grid fabrication technology will be assessed.

- Orbit/radiation tradeoffs will be assessed. There may be significant advantages in a Sun-synchronous, twilight orbit.

Points of Contact:

    Science Panel: Dr. Hugh Hudson (619) 534-4476
                                     CASS
                                       University of California at San Diego
                                     C-011
                                     La Jolla, CA 92093
    Discipline: Solar Physics

Section 6 Candidate Future Missions

Program Manager: TBD
Program Scientist: Dr. Dave Bohlin (202) 453-1514
NASA Center: GSFC
Project Manager: Steve Paddack (301) 344-4879
Project Scientist: TBD

Section 6 Candidate Future Missions

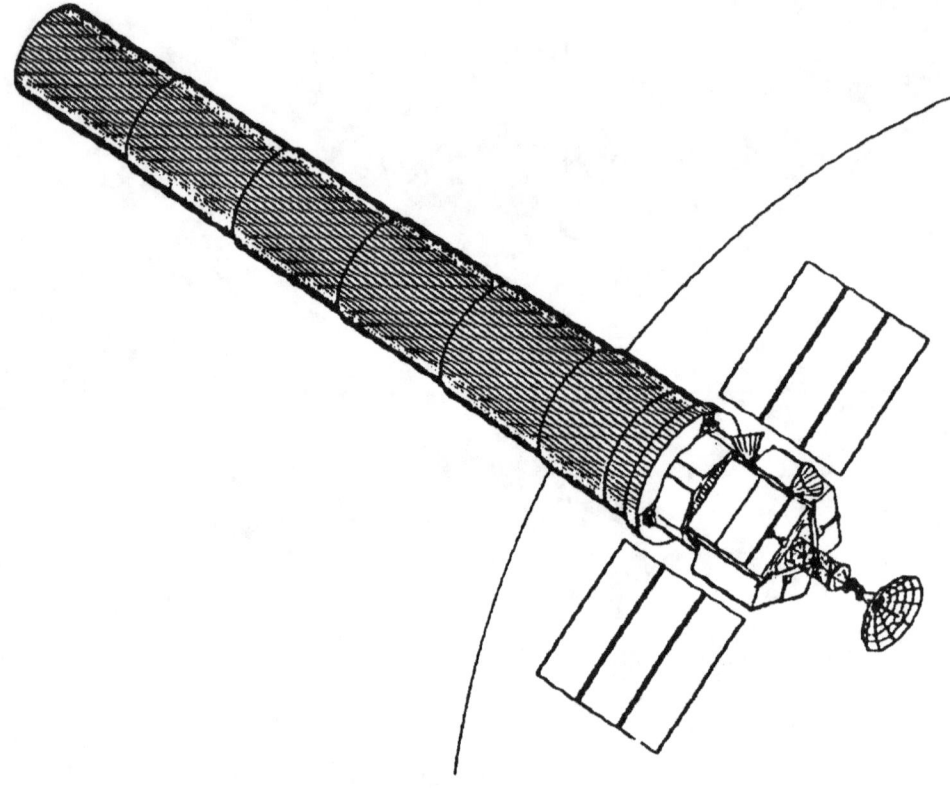

High-Energy Solar Physics

Section 6 Candidate Future Missions

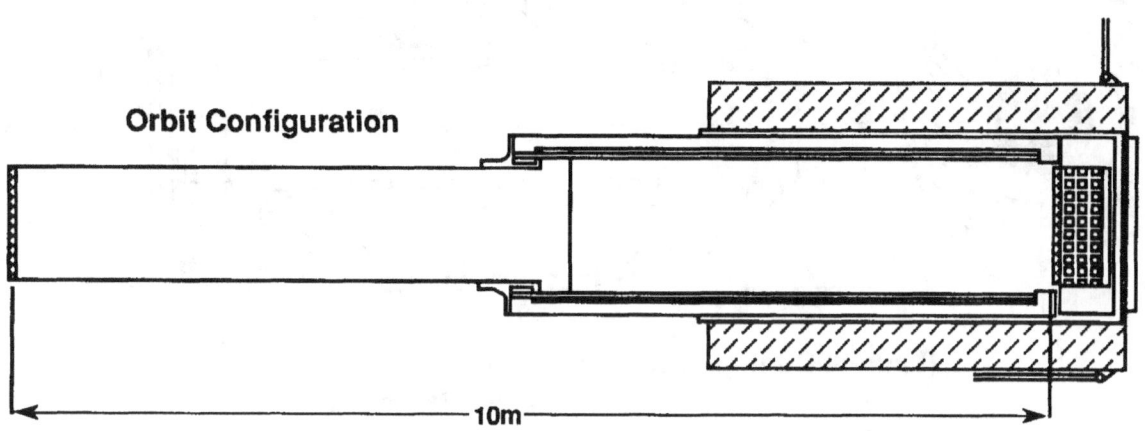

High Energy Solar Physics Mission in orbit

Section 6 Candidate Future Missions

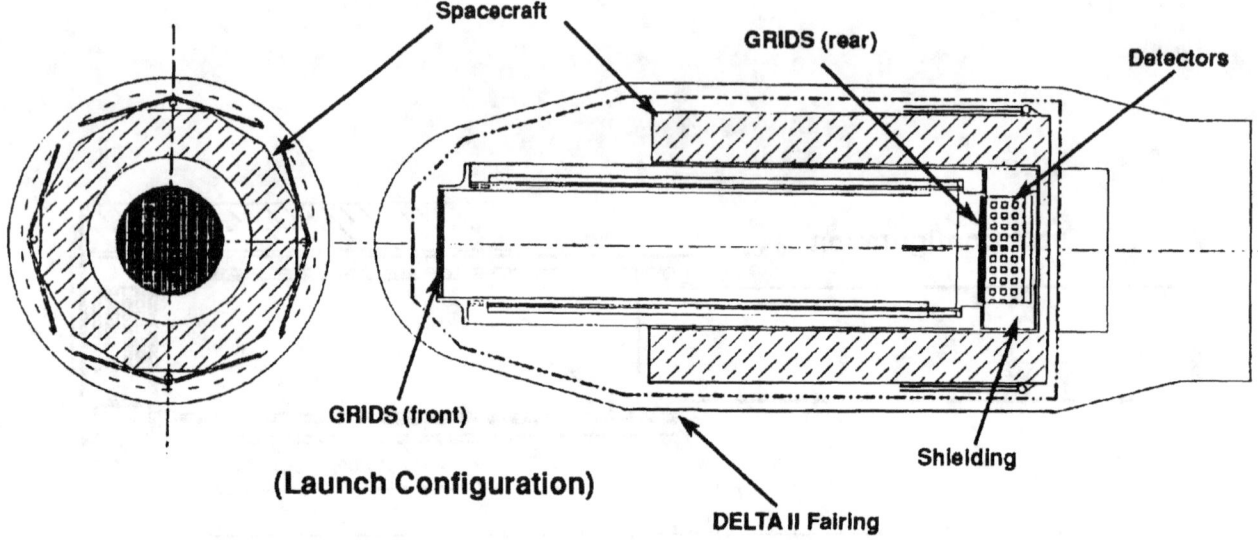

High Energy Solar Physics Mission at launch

- ≤ 28 DEGREE LEO
- SPACECRAFT POINTED AT SUN 90% OF TIME (EXCLUDING ECLIPSES)
- FRONT & REAR GRIDS SEPARATED BY 10 METERS

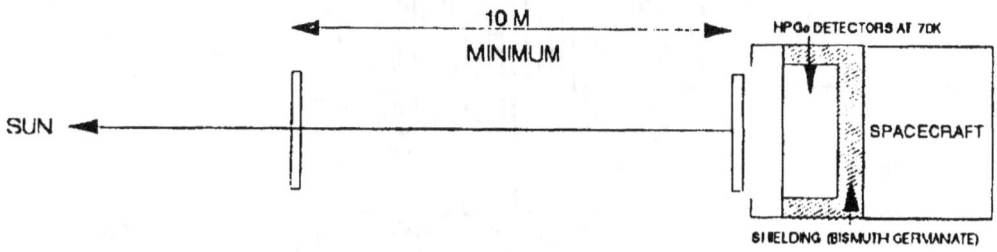

- FOURIER IMAGING REQUIRES ROTATION ABOUT OPTICAL (LONG) AXIS AT 15 RPM
- HIGH-PURITY GERMANIUM (HPGe) DETECTORS (30) COOLED TO ~ 70K°

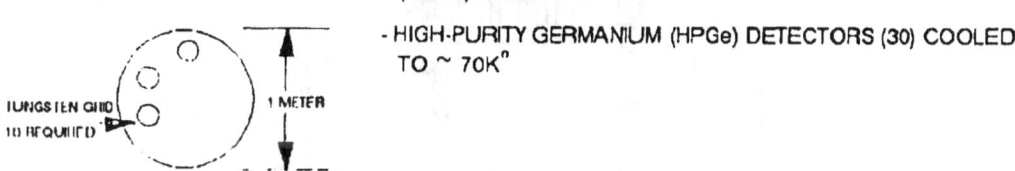

FRONT & REAR GIRD SLITS EACH APPROXIMATELY 150KG
TOTAL PAYLOAD (INSTRUMENT) WEIGHT 1500KG

High-Energy Solar Physics Mission

Section 6 Candidate Future Missions

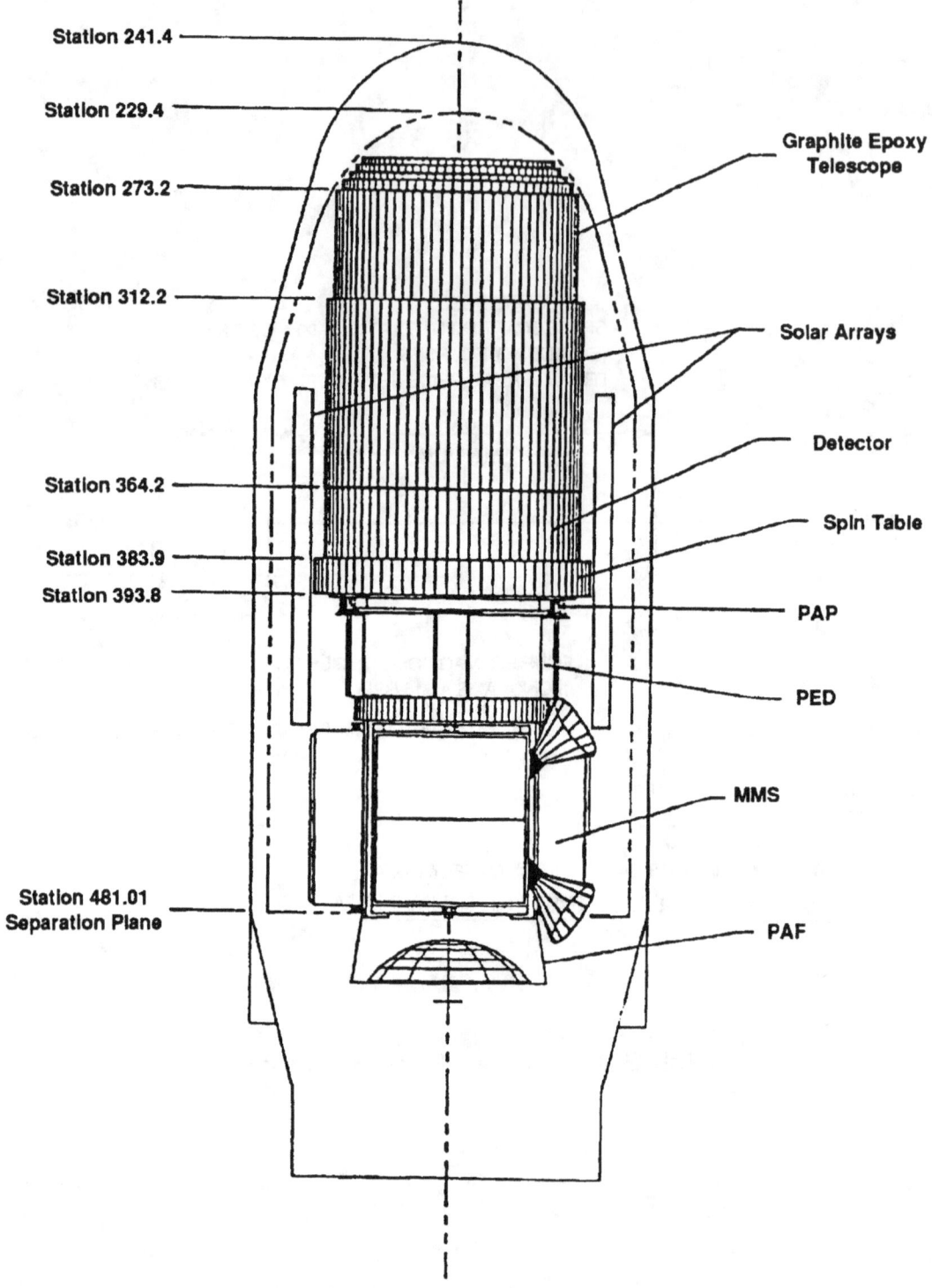

HESP Delta II configuration

Section 6 Candidate Future Missions

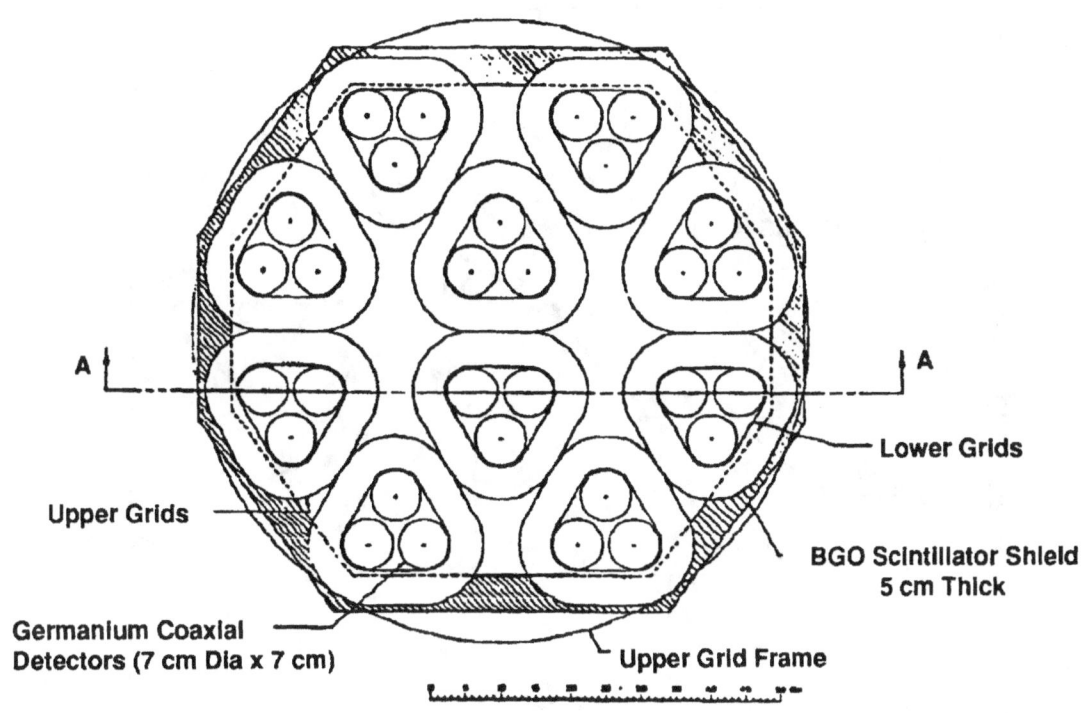

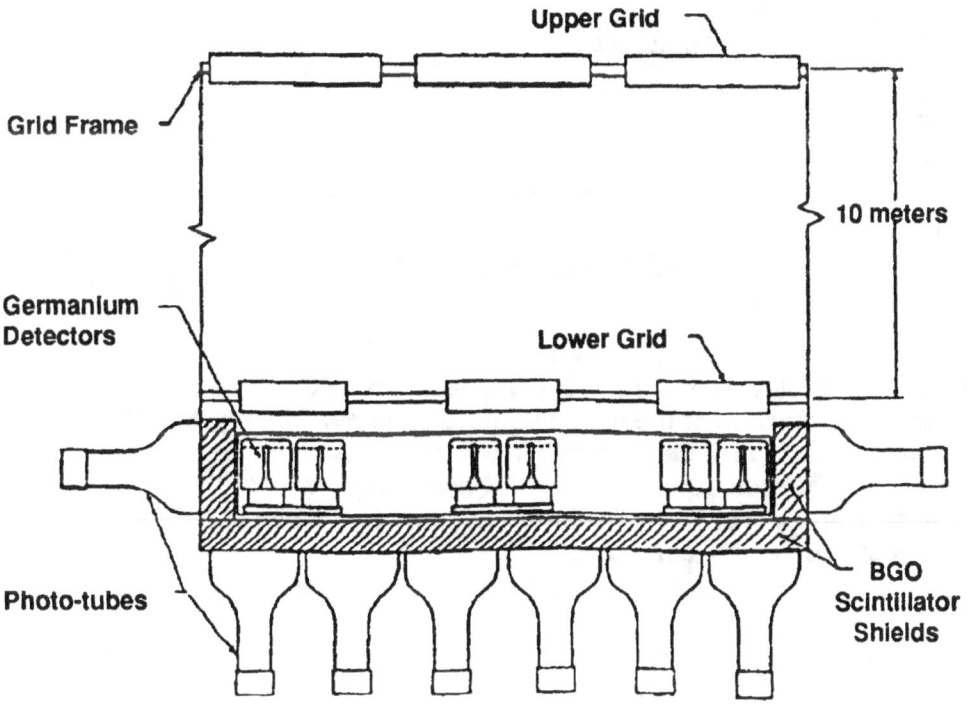

Schematic illustration of the basic components of HEISPEC

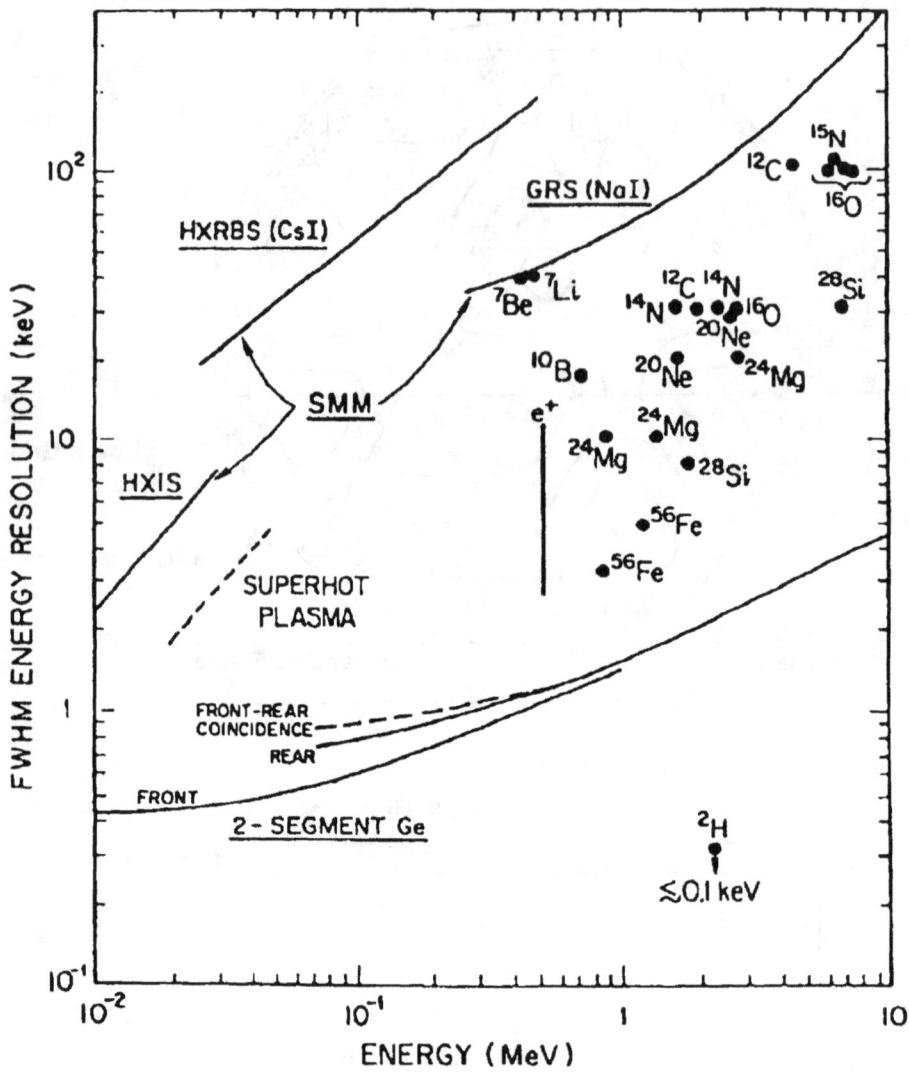

The spectral resolution of the HEISPEC 2-segment HPGe detectors as a function of the incident photon energy. The typical line widths expected for gamma-rays in solar flares are shown together with the resolution of the hard X-ray and gamma-ray instruments on SMM and the resolution required to resolve the steep super-hot thermal component of flares.

Section 6 Candidate Future Missions

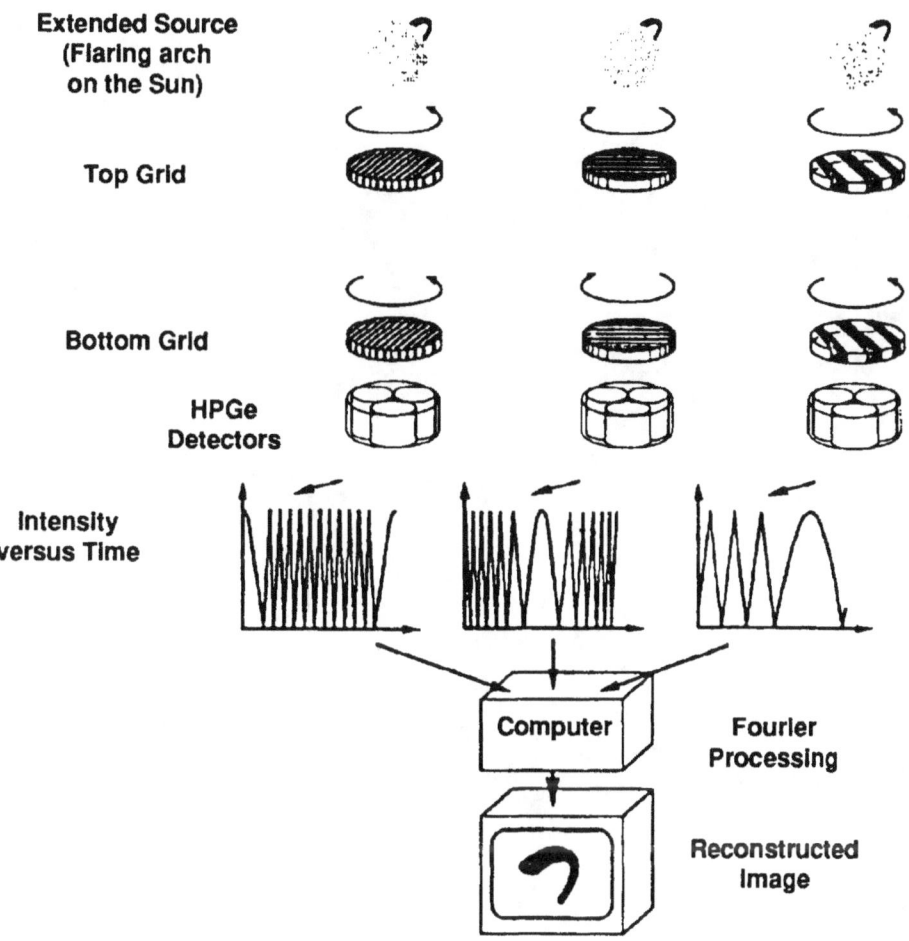

ROTATING MODULATION COLLIMATORS

6.12 Solar Probe

Target: Solar corona

Orbit: Interplanetary EJGA
Perihelion: 4 R_S
Aphelion: AU
Inclination (solar): 90°

Mission Duration: ~9 years (2 encounters)

Mission Class: Major

Mass: TBD

Launch Vehicle: Titan IV (SRMU)/Centaur (November 2000)

Theme: Cosmic & heliospheric physics

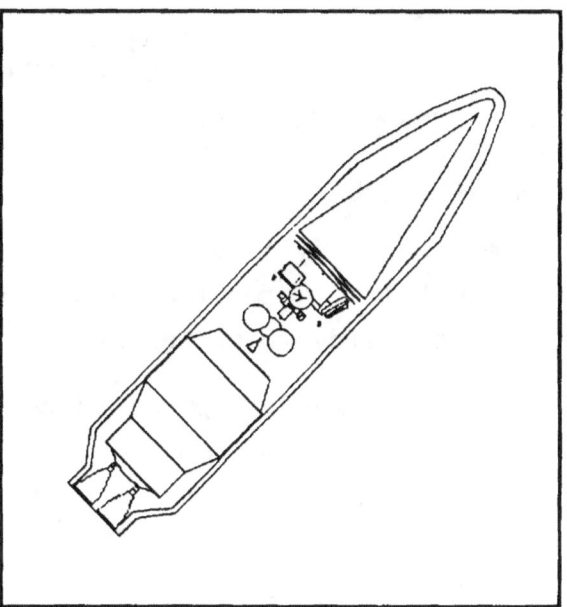

Science Objectives:

- Investigate the inner heliosphere (< 65 R_s)
    - Last major unprobed region of our solar system
    - Fill in gaps in knowledge of extended solar corona

- Coronal structure
    - Large- and small-scale structure
    - Time variations

- Coronal heating
    - Energy transport in lower atmosphere

- Solar-wind acceleration

- Plasma turbulence
    - Within the coronal envelope

- Energetic particles
    - Acceleration, storage and transport

- Interplanetary dust
    - Sources, sinks and dynamics

Section 6 Candidate Future Missions

Spacecraft:

    Type: 3-axis stabilized

    Special Features: Low sublimation heat shield
    Deployable magnetometer boom
    Spinning instrument platform (60 rpm)

    Special Requirements: Ability to withstand environments at 4 R_S

Instruments:

Measurement objectives:

    Fully define the particle and field environment of the directly sampled solar corona
    Spatial and temporal structure

- Plasma package (fast plasma, 3-D ions, 3-D electrons):
  — Measure bulk plasma flow
    - Radial dependence, local turbulence, magnetic topology, spatial and temporal variations, suprathermal ion/electron populations

- Magnetometer experiment:
  — Measure DC magnetic fields and hydromagnetic waves
    - Bulk structure of solar corona

- Plasma wave:
  — Measure ELF/VLF electromagnetic and electrostatic wave turbulence
    - Nature of coronal plasma turbulence

- Thermal and suprathermal ion composition package (TIMCA, SIMCA):
  — Local thermal and suprathermal ion populations
    - Solar wind heavy ion acceleration

- Energetic particles (MEPA, HEPA):
  — Local suprathermal and energetic particle population
    - Transient energy releases, mag topology, coronal turbulence, local spatial coronal filaments

- Coronal spectra imager:
  — 3-D overview of coronal structure
    - Global context and time variability

- Neutron/gamma-ray:
    - Nonlocal probe of solar transients
        - Ion acceleration mechanisms, occurrence rate of solar transients
- Dust experiment:
    - Measure meteoroid flux, mass distribution, and velocity
        - Sources and sinks of zodiacal dust cloud
        - Regulatory mechanisms of dust dynamics

Instrument	Mass (kg)	Power (W)	Data rate (kbps)	Data storage	FOV
Fast Plasma	8	8	5	2.5 Gbits	1 sr (about sensor normal)
3-D Ions	12.5	17	10	5 Gbits	6°x330°
3-D Electrons	3.5	3	5	2 Gbits	15°x330°
Magnetometer	5.5	6	2.5	1.4 Gbits	N/A
Plasma Wave	19	11	15	7.5 Gbits	N/A
Thermal Composition	12	8	5	Limited, temp.	7°x100°
Suprathermal Composition	13	9	10	Limited, temp.	S7°x150°
Medium Energy Particles	10	8	10	Limited, temp.	15°x120°
High-Energy Particles	10	8	1	Limited, temp.	15°x120°
Coronal Spectral Imager	25	15	5	1.5 Gbits	90°x140°
Neutron/Gamma Ray	10	5	1	None	(obstruct of Sun < 10 g/cm2)
Coronal Dust	5	5	0.5	1 Mbit	(delta theta =140°)
TOTAL	133.5	103	70		

Mission Strategy:

The Solar Probe mission would carry out the first *in situ* exploration of the solar corona, penetrating to a height of about 3 solar radii above the photosphere, and is a key mission of the space physics program to study the source of the solar wind plasma and fields. In order to achieve solar orbit the spacecraft could be launched on a 2 AU aphelion trajectory where a propulsive maneuver would retarget for an Earth swingby gravity assist. The resulting trajectory would be toward the planet Jupiter, where Jovian gravity assist would retarget the spacecraft toward the Sun. The goal is to place the spacecraft in an orbit with a perihelion of not more than 4 solar radii, and an inclination of 90 degrees. The primary encounter phase of the mission would be within the orbit of Mercury, and would last about ten days. Pole to pole coverage of the Sun would take about 14 hours.

- Trajectory: ΔV-EJGA
    - Provides proper Earth position @encounter, C_3
    - Trajectory correction maneuvers (TCMs) executed during cruise and at Jupiter approach
- Jupiter Gravity Assist (JGA): March 2004
    - Front side retrograde flyby
    - ± 30 day encounter period @Jupiter
    - ~90 min solar and Earth occultation @Jupiter closest approach
    - TCMs executed post flyby for cleanup and initial perihelion targeting
- Perihelion 1: June 2004
    - Primary mission ± 5 days (inside 65 R_S)
    - Primary data acquisition ± 5 hours
    - Total flight time P1~5.5 years
    - Final period trimmed to 2.5 years, maneuver executed outside ± 5 hours
    - Real-time @ 70 kb/s, record @ ~ 190 kb/s
- Perihelion 2: March 2008

Technology Requirements:

- Low-sublimation heat shield—3000 Suns (2100°k)
  <2.5 mg/sec mass loss from 200kg shield
- High bit-rate communications (ka band) through Solar Corona
  70 kb/sec at Perihelion
- Spacecraft power source (new RTGs)
- Refractory antenna for plasma wave experiment
- Temperatures necessitate hybrid materials design

Points of Contact:

Science Panel:	Dr. Thomas E. Holzer (303) 497-1536
	High Altitude Observatory
	National Center for Atmospheric Research
	P.O. Box 3000
	Boulder, CO 80307
Discipline:	Cosmic and heliospheric physics
Program Manager:	TBD
Program Scientist:	Dr. Vernon Jones (202) 453-1514
NASA Center:	JPL
Project Manager:	Ron Boain (818) 354-5122
Project Scientist:	TBD

Section 6 Candidate Future Missions

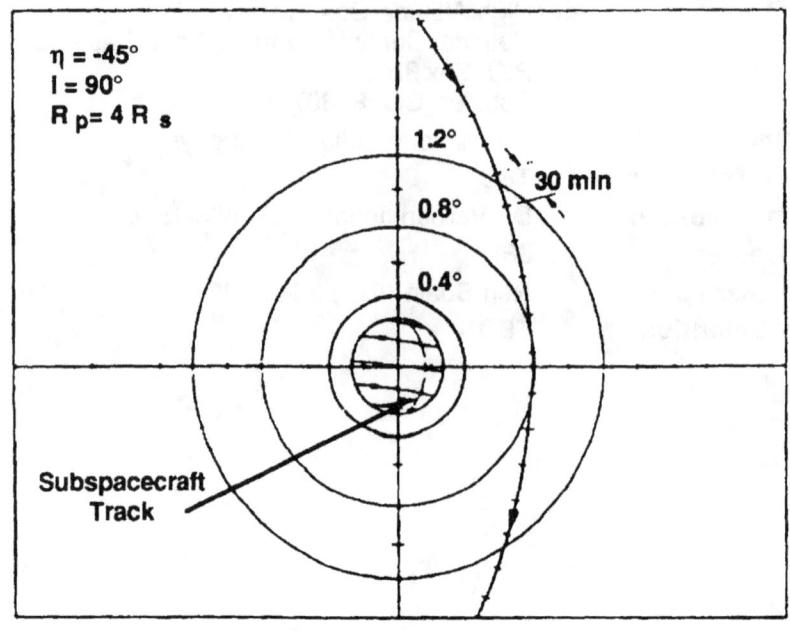

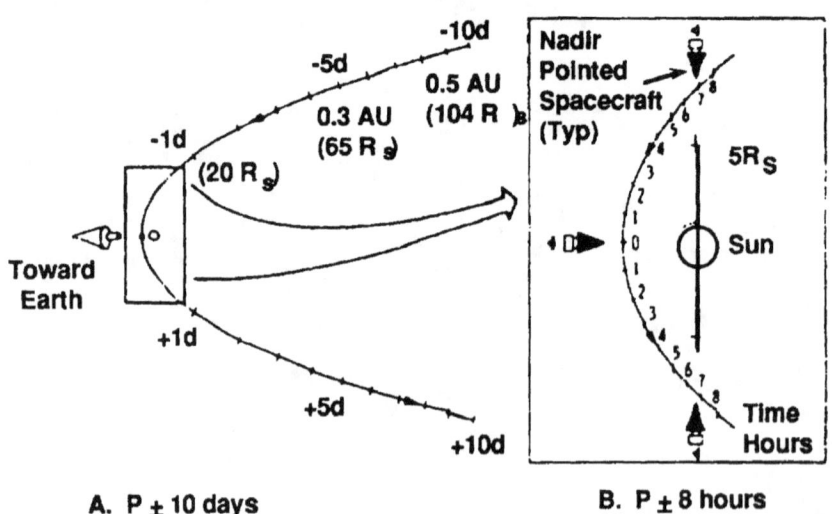

A. P ± 10 days

B. P ± 8 hours

Solar Probe orbit

Section 6 Candidate Future Missions

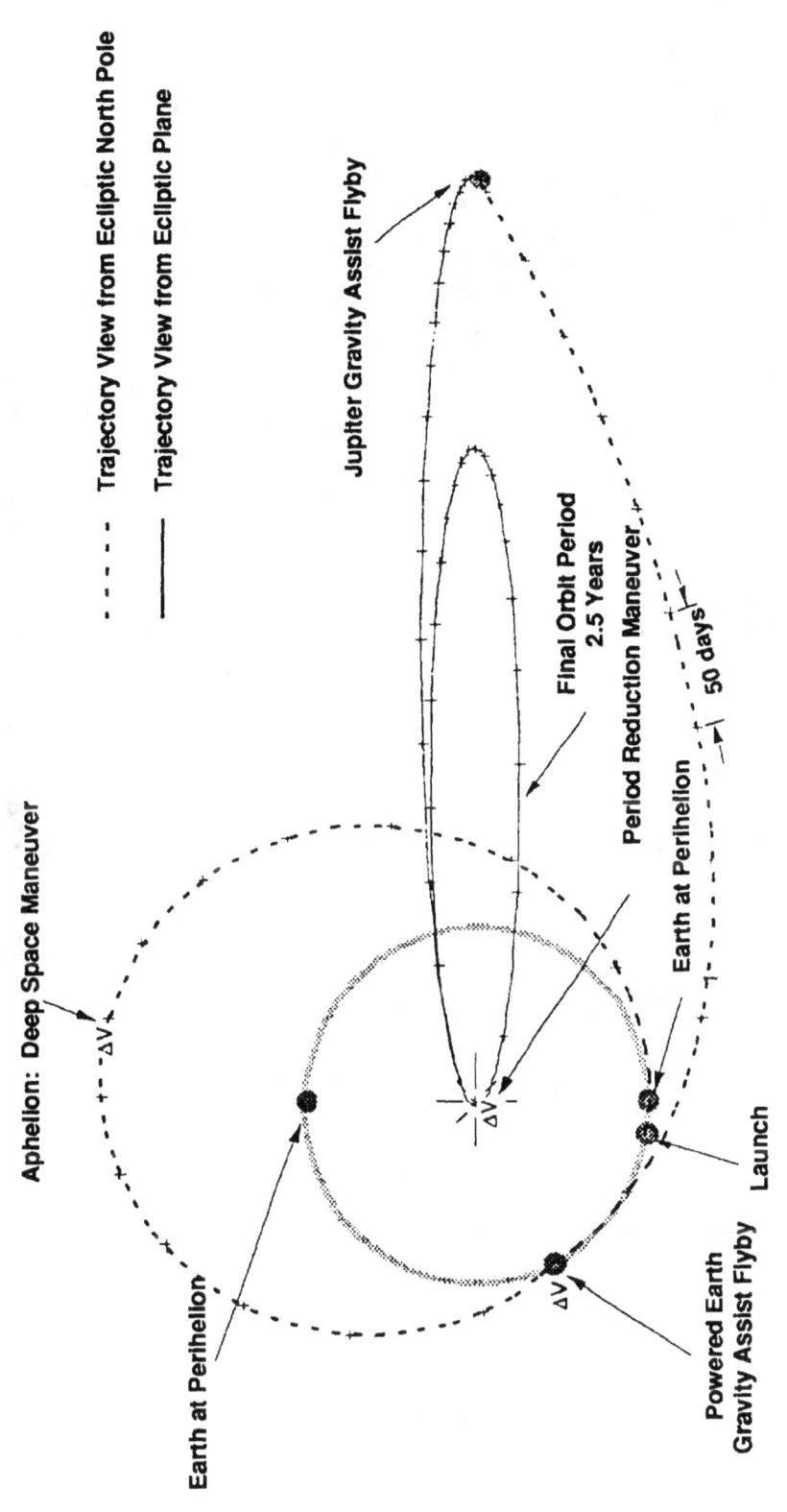

Solar Probe technical and programmatic review

Section 6 Candidate Future Missions

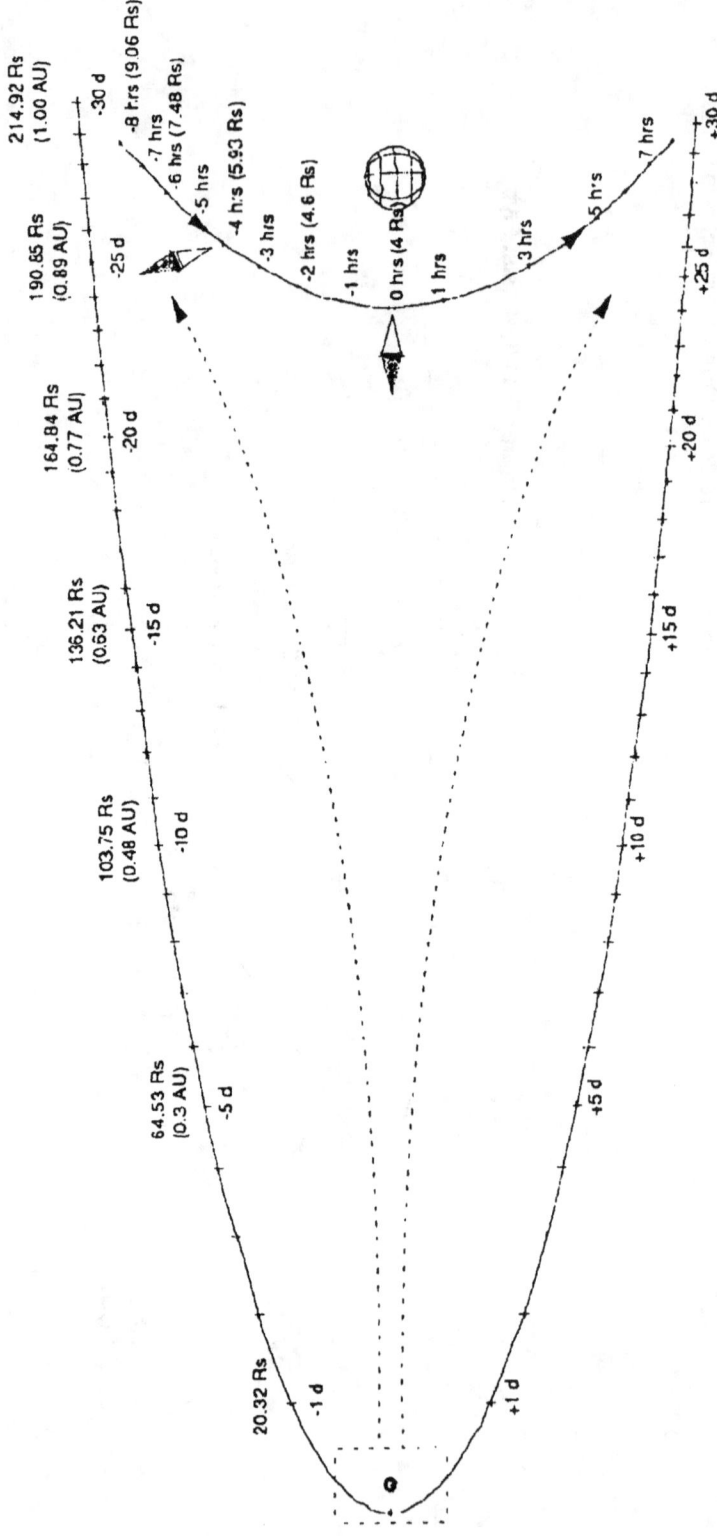

Solar Probe near perihelion reference trajectory

Section 6 Candidate Future Missions

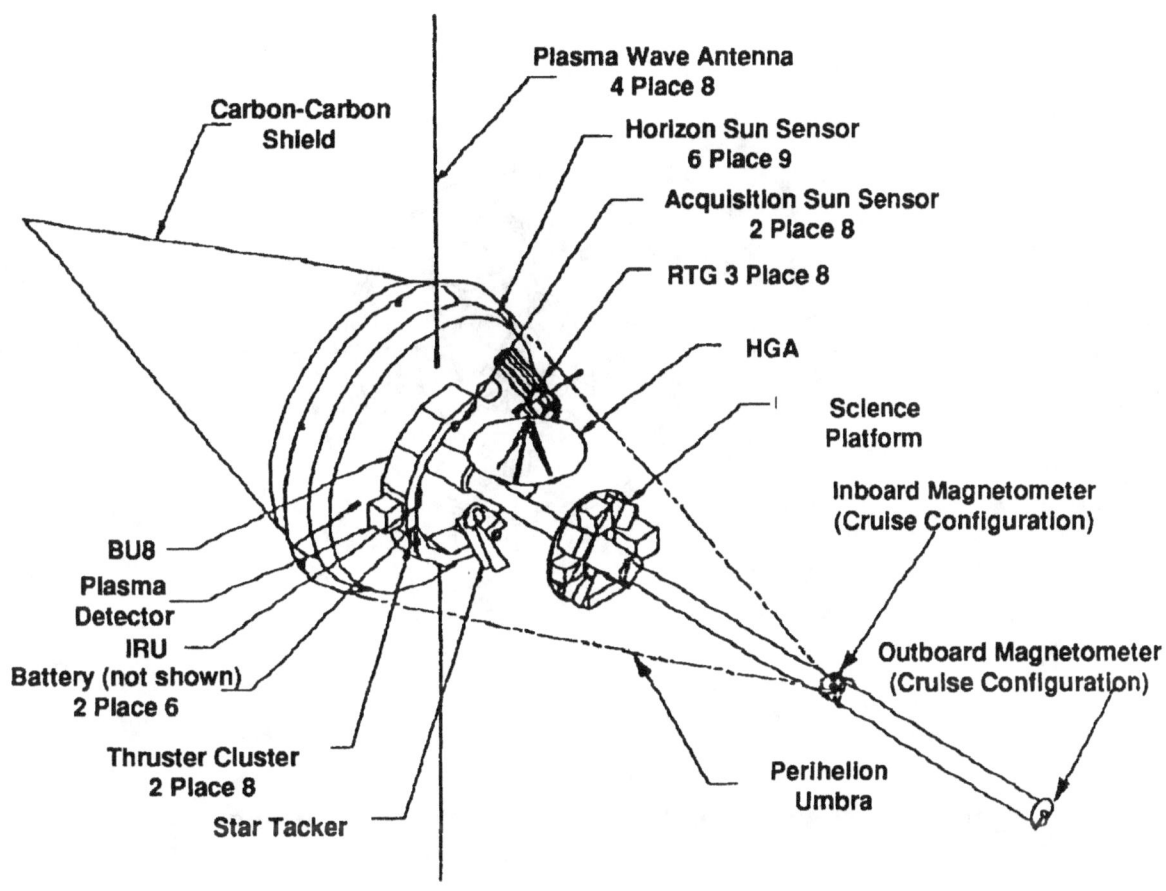

Solar Probe spacecraft

Section 6 Candidate Future Missions

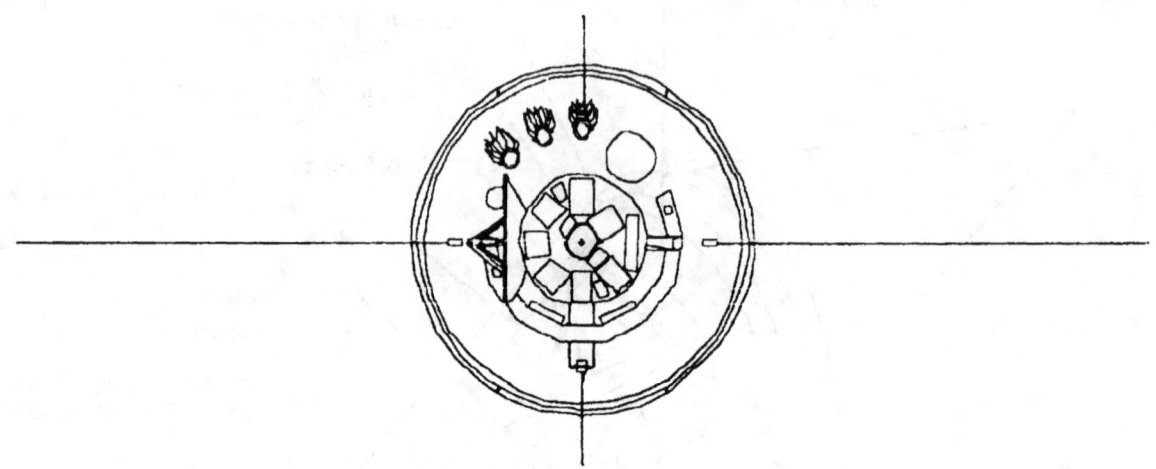

Baseline 3-axis stabilized spacecraft configuration—top view

Section 6 Candidate Future Missions

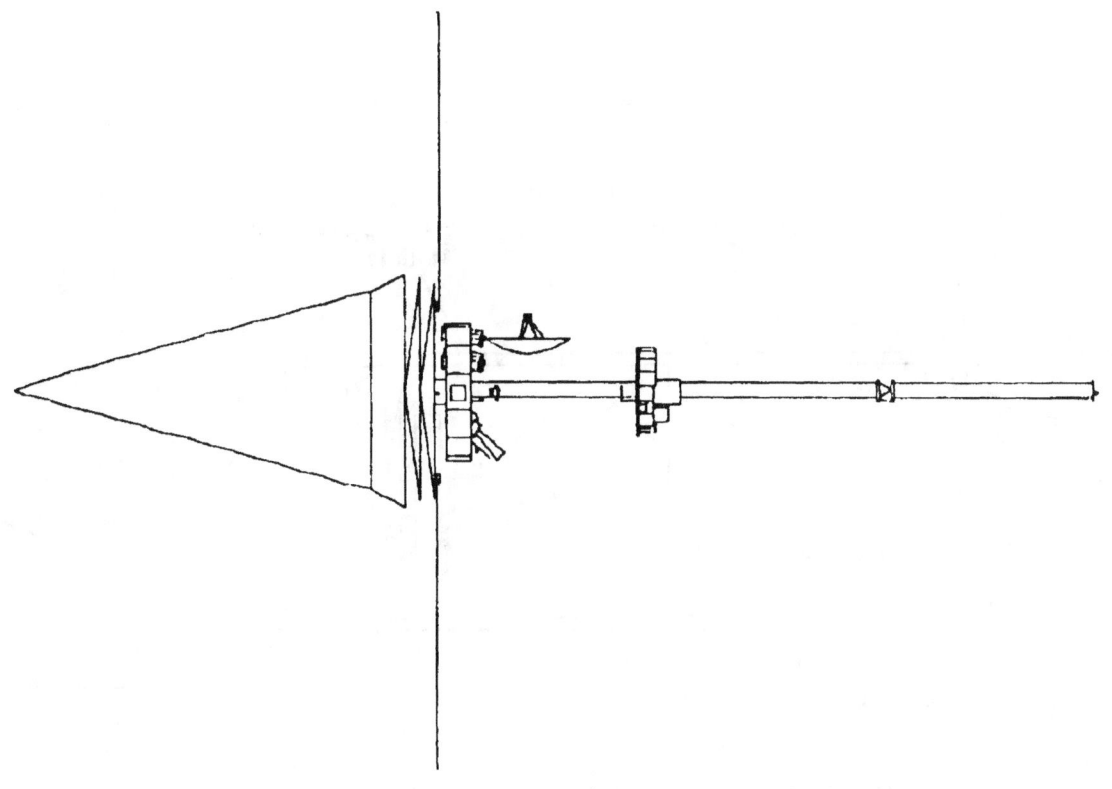

Baseline 3-axis stabilized spacecraft configuration—side view

Section 6 Candidate Future Missions

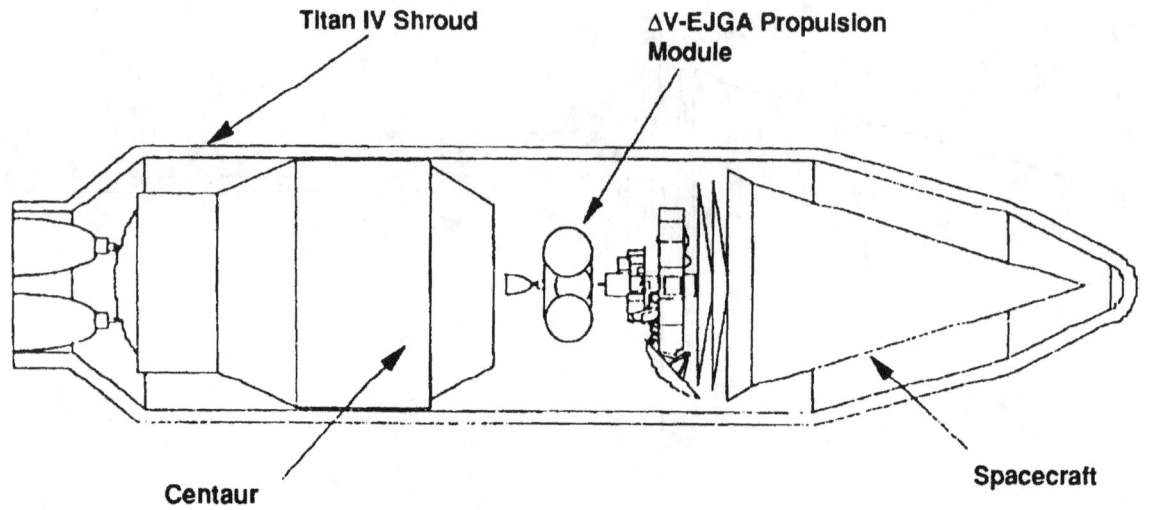

ΔV-EJGA spacecraft launch configuration on the Titan IV/Centaur vehicle

Section 6 Candidate Future Missions

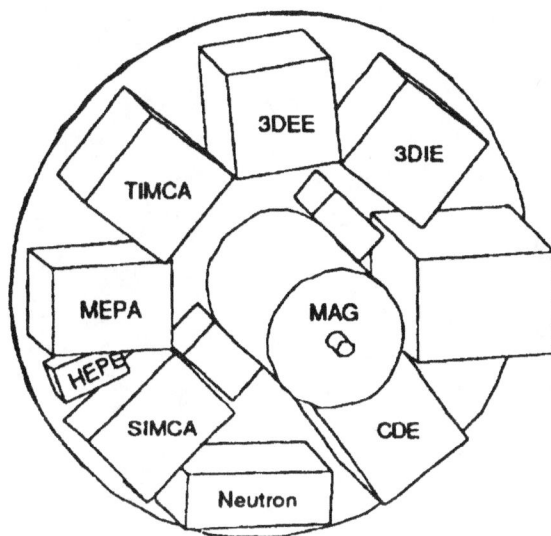

3DIE : 3-D Plasma Ion Experiment

3DEE : 3-D Plasma Electron Experiment

MAG : Magnetometer

TIMCA : Thermal Ion Mass and Charge Analyzer

SIMCA : Suprathermal Ion Mass and Charge Analyzer

MEPA : Medium Energy Particle Analyzer

HEPE : High Energy Particle Experiment

Neutron : Neutron/Gamma Ray Experiment

CDE : Coronal Dust Experiment

Solar Probe instrument pallet

6.13 Polar Heliosphere Probe

Target: Distant polar heliosphere (to ≥ 15 AU)

Orbit: ~Normal to ecliptic, over the Sun's pole to the outer heliosphere
Solar latitude: 70° to 90°
Aphelion: 12 AU

Mission Duration: 10–15 years

Mission Class: Moderate

Mass: 367 kg
545 kg at injection
55 kg payload

Launch Vehicle: Atlas -II AS (or Titan IV/Centaur)

Theme: The next frontier: the global heliosphere & interstellar space

Science Objectives:

To explore the distant polar heliosphere, in order to study new aspects of large-scale solar wind structure and evolution, energetic particle acceleration and transport, solar wind interaction with the interstellar medium, and fundamental plasma processes.

Spacecraft:

Type: Spinner (Ulysses class)
Special Features: Standard magnetometer boom and plasma antenna, etc.
Special Requirements: None

Instruments:

	Mass (kg)	Power (W)	Data rate (bps)	Data storage	FOV
Magnetometer	5	5	80	TBD	TBD
Solar wind analyzer	13	10	248	TBD	TBD
Suprathermal ions & electrons	6	4	160	TBD	TBD
Cosmic Ray Analyzer	19	18	176	TBD	TBD
Plasma waves	7	11	232	TBD	TBD
UV photometer	5	5	104	TBD	TBD
Total	55	53	1000	store & dump	

Note: The instruments listed here (with the exception of the UV Photometer) are from Ulysses, not because they are *the* instruments to use, but because they are typical of the class of instrument desired. Several of the instruments could also be from Voyager, for instance.

Magnetometer:

　Measurement Objectives: The distant high-latitude solar magnetic field.

　Principle of Operation: Dual magnetometer system.

　Accommodation and Integration Requirements: The magnetometer should be mounted on a magnetically clean boom, as protection from spacecraft-generated fields. Spacecraft magnetic cleanliness is important.

　Technology Readiness: No technology development needed.

　Design Heritage: Ulysses and Voyager Magnetic-Field Investigation.

Solar Wind Analyzer (SWA):

　Measurement Objectives: The distant high-latitude solar wind speed and density, proton and electron temperatures, and composition (H through Fe).

　Principle of Operation: Spacecraft rotation (5RPM) is used to sample particles coming from different clock angle directions. Plasma measurements at different cone angles are performed by the instrument. Composition is measured with an electrostatic analyzer with post-acceleration, followed by time-of-flight and energy measurement.

Accommodation and Integration Requirements: The SWA requires an unobstructed view of the solar wind ±73° from the normal to the spin axis.

Technology Readiness: No technology development needed.

Design Heritage: Ulysses or Voyager Solar-Wind Plasma Experiment and Ulysses Solar-Wind Ion Composition Spectrometer.

Suprathermal Ions and Electrons (SIE):

Measurement Objectives: Spectra of ions > 50 keV energy and of electrons > 30 keV

Principle of Operation: Several similar solid-state detector telescopes give essentially complete pitch angle coverage.

Accommodation and Integration Requirements: The Ulysses SIE requires an unobstructed view of the solar wind ±82.5° from the normal to the spin axis.

Technology Readiness: No technology development needed.

Design Heritage: Ulysses Energetic Particle Composition Experiment, Voyager LEPC.

Cosmic-Ray and Solar Particle Analyzer (CRA):

Measurement Objectives: Solar modulation of cosmic rays, and the source, confinement, and propagation of cosmic-rays.

Principle of Operation: Several quite distinct telescope subsystems are used to cover elements from H to Ni over the energy range ~5–1000 MeV/nucleon.

Accommodation and Integration Requirements: The Ulysses CRA requires access to the ambient environment ±45° from the normal to the spin axis.

Technology Readiness: No technology development needed.

Design Heritage: Ulysses Cosmic-Ray and Solar Particle Investigation, Voyager CRS.

Radio & Plasma Waves (RPW):

Measurement Objectives: Both remote sensing of the heliosphere via observation of distant radio sources, and *in-situ* sensing of local wave phenomena.

Principle of Operation: Two 35-m antennas are used as a single 72.5-m dipole for electric wave reception over 1 kHz to 1 Mhz. Sounding is performed by briefly using

Section 6 Candidate Future Missions

the antenna to transmit and then observing resonance phenomena. A two-axis search coil is used as a magnetic field antenna.

Accommodation and Integration Requirements: Electrical/magnetic cleanliness is crucial.

Technology Readiness: No technology development needed.

Design Heritage: Ulysses Unified Radio and Plasma Wave Experiment.

UV Photometer (UVP):

Measurement Objectives: The column density of neutral H and He atoms.

Principle of Operation: The UVP measures the brightness of the back-scattered solar UV radiation in various directions. The brightness is a function of the amount of neutral material along the line-of-sight. By combining observations from many different geometries a 3-D map of the density of neutral particles can be made.

Accommodation and Integration Requirements: TBD.

Technology Readiness: No technology development needed.

Design Heritage: Pioneer 10,11.

Mission Strategy:

The baseline trajectory is a 5-year ΔVEGA with a deep-space ΔV of roughly 0.8 km/s, followed by a Jupiter gravity assist with a flyby altitude of about 5.6 R_J. This gives an inclination of 78° and an aphelion of 12 AU. The heliospheric flythrough could be improved to 90° and 20 AU by switching to a 4-year ΔVEGA with a ΔV of roughly 2 km/s. The tradeoff between the resultant increase in launch mass and the gaining of the highest-latitude data remains for a future study.

- Atlas-IIAS can deliver 20 AU, 90° polar heliosphere probe
- Use (2 yr + 5 yr) ΔVEGA + JGA trajectory
- 20 years to reach target
- Target to 20 AU, 90°
- Difference to be made by using Titan IV/Centaur is 6 years in flight time

Spacecraft System Description:

System Design Concept: PHP is a spinning spacecraft, like Ulysses (the baseline) or Pioneer 10/11. The spin axis is maintained toward Earth for communications, with power being supplied by an RTG. The total mass is 370 kg (exclusive of the solid rocket propulsion subsystem for the deep-space ΔV), of which about 60 kg is the scientific payload. Minimization of electromagnetic noise and magnetic fields is a high priority.

Subsystem Descriptions:

Altitude Control Subsystem: Sun sensors and high gain antenna pointing error. Passive fluid-in-tube nutation damper.

Power subsytem: RTG (250–300 W) with external and internal shunts.

Propulsion Subsystem: Monoprop hydrazine subsystem for navigation and attitude control (150 m/s capability). It is assumed that the Ulysses subsystem will suffice. A solid rocket motor is added for the ΔVEGA deep-space burn (0.8 km/s).

Structure and Mechanisms Subsystem: Rectangular box with two overhanging "balconies." Two deployable wire antennas, each 35 m long; one deployable monopole antenna, 7.5 m; one double-hinged boom, 5.6 m.

Command and Data Handling Subsystem: Telemetry rates from 64 to 8192 b/s. Onboard storage of 45 Mbits.

Thermal Control Subsystem: Radiator on one side of spacecraft, with rest of spacecraft superinsulated.

Communications Subsystem: 1.65-m parabolic High Gain Antenna, 20-W X-band TWT, 5-W solid-state S-band amplifier.

Technology Requirements: None

Points of Contact:

Science Panel:	Dr. Leonard F. Burlaga (301) 286-5956 Interplanetary Physics Branch, Code 692 Goddard Space Flight Center National Aeronautics and Space Administration Greenbelt, MD 20771
Discipline:	Cosmic and heliospheric physics
Program Manager:	TBD
Program Scientist:	Dr. Vernon Jones (202) 453-1514
NASA Center:	JPL: Project, GSFC: Science
Project Manager:	Ron Boain (818) 354-5122
Project Scientist:	TBD

Section 6 Candidate Future Missions

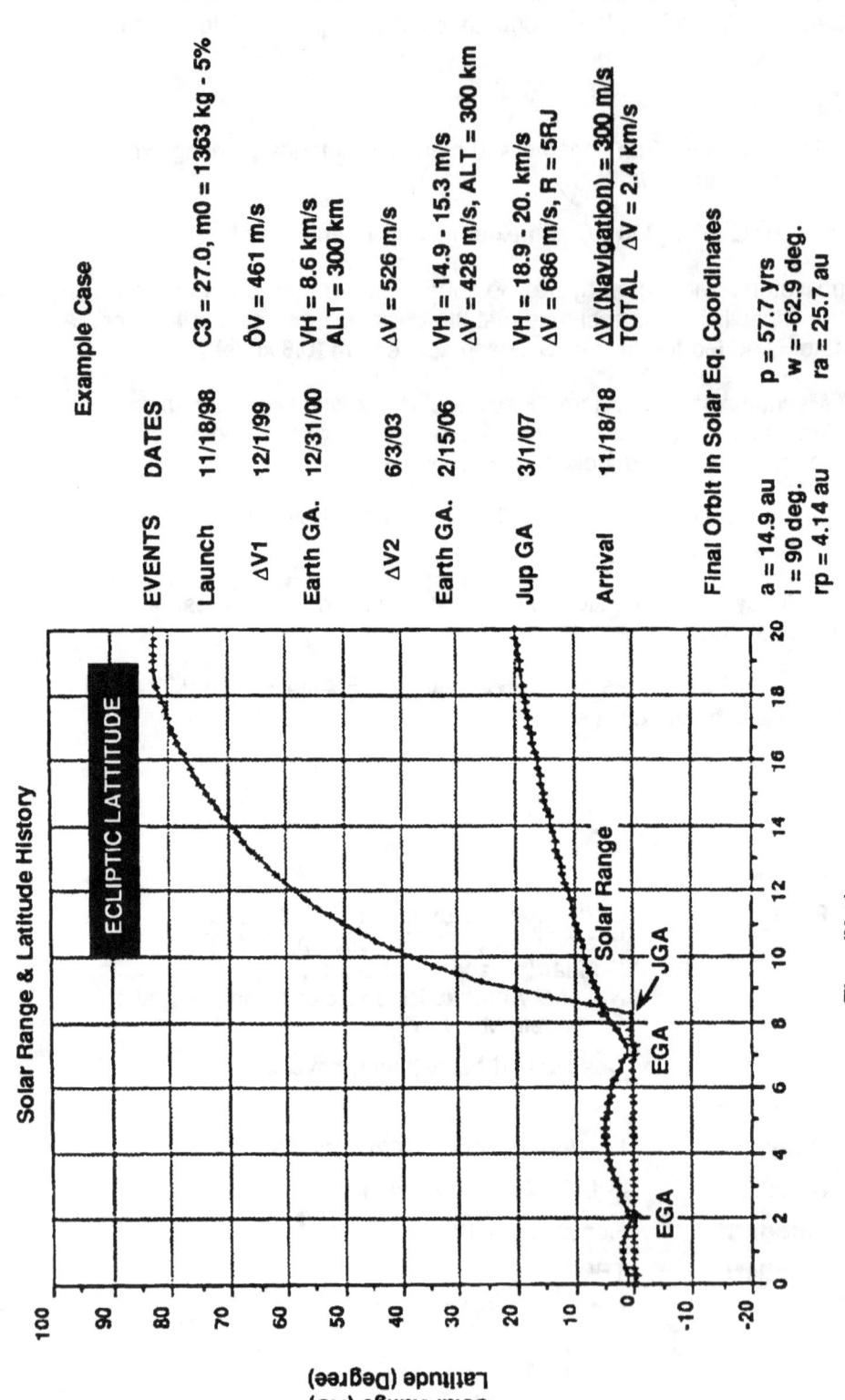

Example Case

EVENTS	DATES	
Launch	11/18/98	C3 = 27.0, m0 = 1363 kg - 5%
ΔV1	12/1/99	δv = 461 m/s
Earth GA.	12/31/00	VH = 8.6 km/s, ALT = 300 km
ΔV2	6/3/03	ΔV = 526 m/s
Earth GA.	2/15/06	VH = 14.9 - 15.3 m/s, ΔV = 428 m/s, ALT = 300 km
Jup GA	3/1/07	VH = 18.9 - 20. km/s, ΔV = 686 m/s, R = 5RJ
Arrival	11/18/18	ΔV (Navigation) = 300 m/s, TOTAL ΔV = 2.4 km/s

Final Orbit in Solar Eq. Coordinates

a = 14.9 au   p = 57.7 yrs
i = 90 deg.   w = -62.9 deg.
rp = 4.14 au   ra = 25.7 au

1998 Helio-Polar Probe to 20 AU, 90 deg.
2 yr + 5 yr ΔVEGA and JGA trajectory

6.13-6

Section 6 Candidate Future Missions

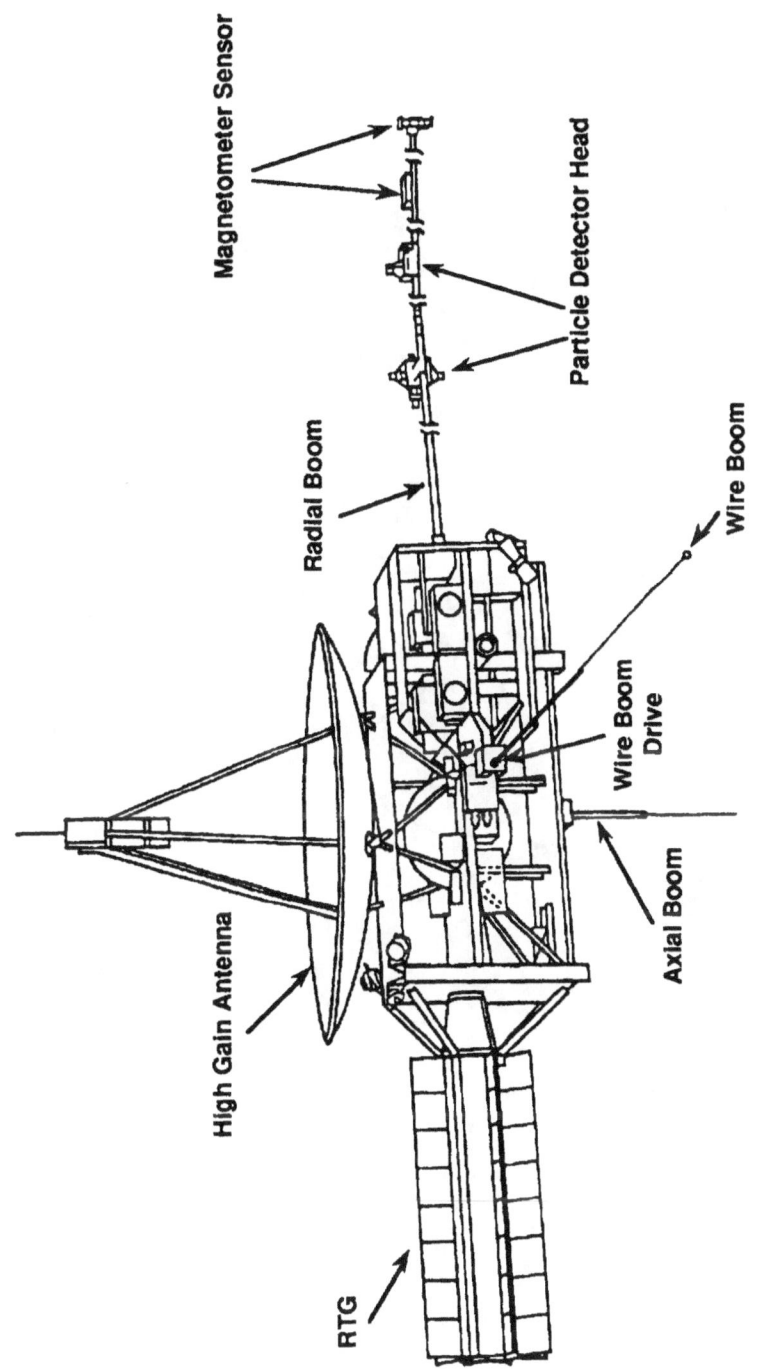

Polar Heliospheric Probe configuration after deep space ΔV

6.14 Interstellar Probe

Target: Heliopause/nearby interstellar medium at 200 AU

Orbit: Jupiter flyby & solar flyby at ~4RS coupled with rocket burn to achieve solar system escape

Mission Duration: 20+ years

Mission Class: Major

Mass: ~100 kg (payload only)

Launch Vehicle: TBD

Theme: The next frontier: the global heliosphere & interstellar space

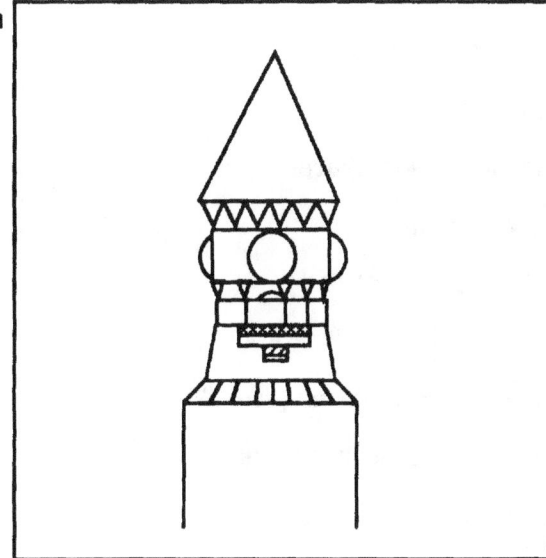

Science Objectives:

Explore the solar wind termination shock, the heliopause, and the nearby interstellar medium. Measure *in situ* acceleration of anomalous cosmic rays. Measure the composition of the nearby interstellar medium and the properties of its plasma, particles, and fields. Measure the spectrum of low-energy cosmic rays outside the heliosphere.

Spacecraft:

Type: Spinner, Pioneer 10, 11/Voyager hybrid design

Special Features: No pointing requirements; magnetometer & 2 RTG booms

Special Requirements: Need to measure 3-D plasma distributions

Instruments:

Key measurements need to be made in the outer heliosphere, in the termination-shock and heliopause boundary regions, and in the nearby interstellar medium. Key parameters to measure in these regions include the plasma density, temperature, flow directions, and composition (charge state, elemental, isotopic); the anomalous and galactic cosmic ray composition, energy spectra, and anisotropy (nuclei and electrons); magnetic and electric field density, direction, and waves; interplanetary and interstellar dust (mass distribution, composition, spectra); and remote sensing and direct measurement of interstellar neutrals.

Section 6 Candidate Future Missions

	Mass (kg)	Power (W)	Data rate/rate @ 200 AU (bps)
Interstellar Neutrals	4	5	16/10
Plasma Composition	15	12	30/40
Magnetometer	3	3	5/1
Plasma/Radio Waves	11	6	800/30
UV Photometer	1	1	10/10
Plasma: Solar Wind	4	4	-/5
Plasma: Interstellar RAM	4	4	-/5
Plasma: Hot Plasma (+DPU)	12	6	20/10
Anomalous Cosmic Rays	8	5	15/15
Galactic Cosmic Rays	15	15	100/25
Electrons and Positrons	10	10	50/30
Suprathermal Ions	6	6	500/30
Suprathermal Protons/Electrons	2	2	-/5
Dust/Composition Detector	8	9	512/10
IR Flux Instrument	2	1	10/10
TOTAL	105	89	2068/236

Mission Strategy:

Achieve as high a terminal velocity as possible, pass through the termination shock and heliopause and penetrate as far as possible into the interstellar medium.

Mission Requirements:

- Reach a distance of 200 AU from the Sun within 25 years of launch.
- Travel in direction of Sun's travel through ISM-270 degrees from Aries, 7 degrees N of ecliptic plane.

Mission/System Options:

Option A:

- Powered Jupiter flyby (ΔVEJGA)
- Shuttle–C/Centaur launch vehicle
- Baseline spacecraft design (1000 kg)
- 38 years to 200 AU (5.71 AU/yr hyperbolic)

Option B:

- Powered Sun flyby (ΔVEJSGA)
- Shuttle–C/Centaur launch vehicle
- Sun shield required (Solar Probe technology)
- Spacecraft design modified to meet packaging constraints (1300 kg w/Sun shield)
- 35.8 years to 200 AU (6.6 AU/yr hyperbolic)

Option C:

- Powered Sun flyby (ΔVEJSGA)
- Titan IV/Centaur launch vehicle
- Sun shield required
- Descoped spacecraft design for ILw mass (825 kg w/Sun shield)
- 34 years to 200 AU (6.98 AU/yr hyperbolic)

Technology Requirements:

Instruments are standard. Need to adapt Solar Probe technology to achieve required velocity, or use some other high propulsion technology.

- Low-thrust propulsion
- 4-solar radii heat shield
- Long range communications—1.5 kbps date return at 200 AU
- Spacecraft energy source
- Spacecraft design life and instrument reliability—35-year system lifetime

Points of Contact:

Science Panel: Dr. Richard A. Mewaldt (818) 356-6612
Mail Stop 220-47
Downs Laboratory
Division of Physics, Mathematics, and Astronomy
California Institute of Technology

Section 6 Candidate Future Missions

Discipline: Cosmic and heliospheric physics
Program Manager: TBD
Program Scientist: Dr. Vernon Jones (202) 453-1514
NASA Center: JPL
Project Manager: Ron Boain (818) 354-5122
Project Scientist: TBD

Section 6 Candidate Future Missions

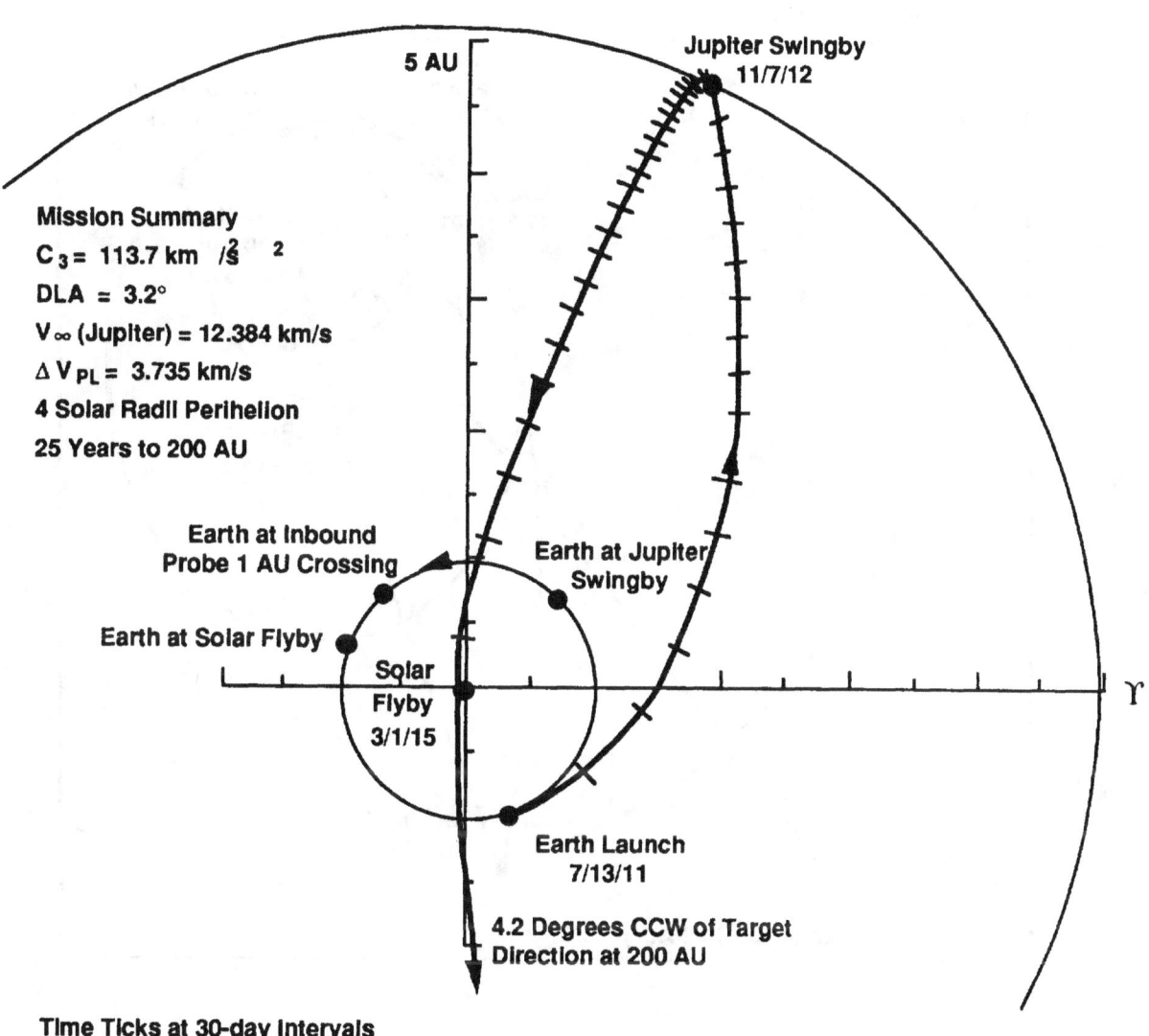

Interstellar Probe 2011 launch trajectory profile

Section 6 Candidate Future Missions

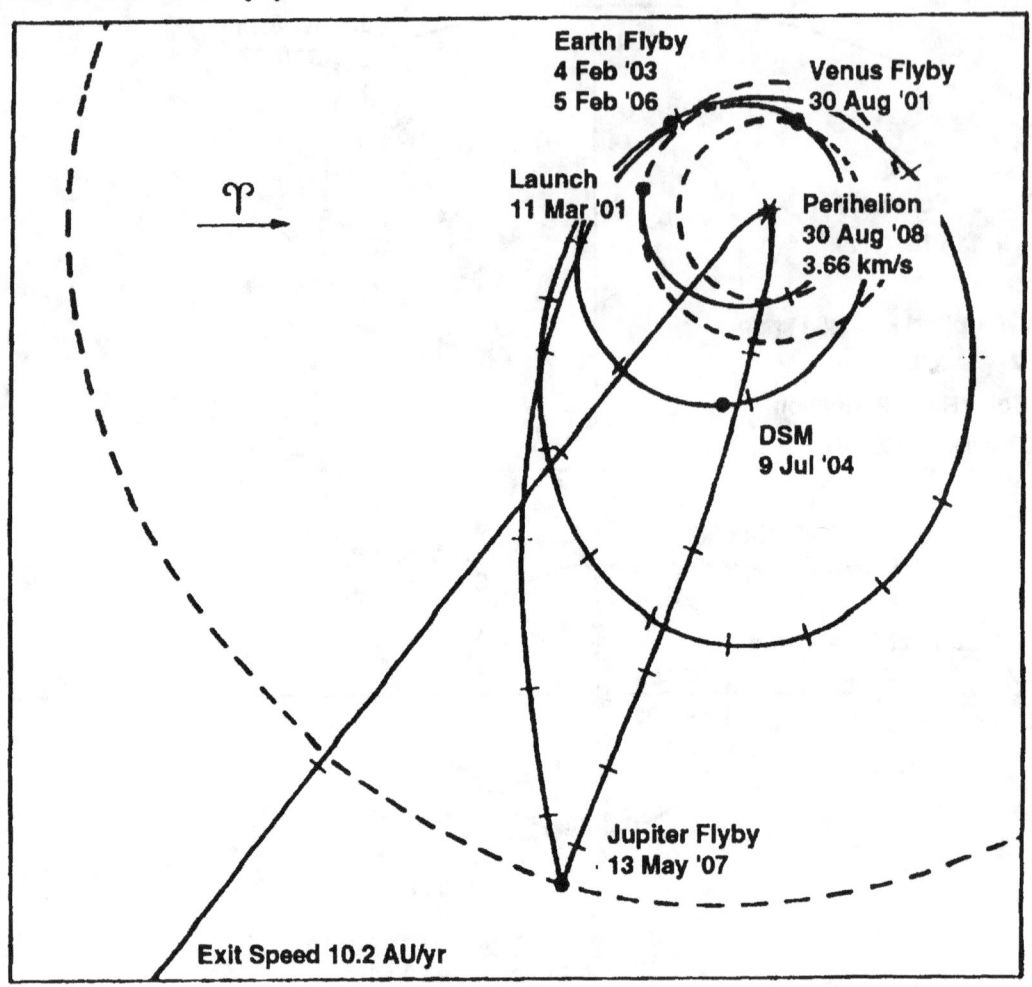

Interstellar Probe

Section 6 Candidate Future Missions

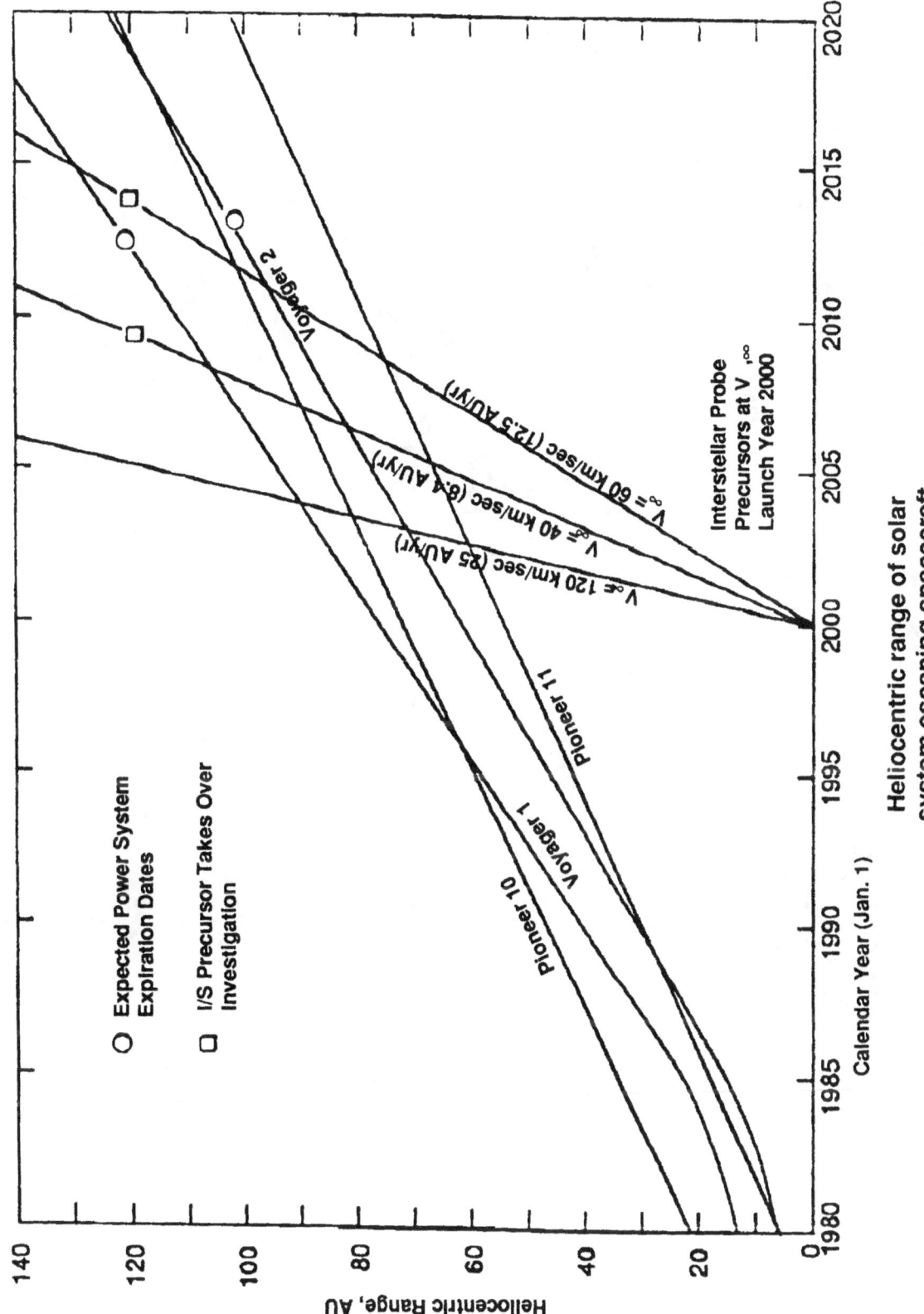

6.14-7

6.15 Time Dependent Global Electrodynamics

Target: Earth's thermosphere/ionosphere

Orbit: 36 S/C: 700 km circular, 4 planes, 90° inclination
1 S/C (EE class): 450 km x 800 km, 90° inclination
2 S/C: 3° either side of EE-class S/C

Mission Duration: 6 months required
2 years desired

Mission Class: Moderate

Mass: Microsats: TBD
EE class: 600 kg at launch

Launch Vehicle: Microsats: Delta II (or Pegasus)
EE-class: Taurus

Theme: Electrodynamic driving forces within the thermosphere/ionosphere

Science Objectives:

Our present understanding of and modeling capabilities within the thermosphere/ionosphere is limited by a lack of knowledge of the electrodynamic driving forces. Understanding and predicting the dynamic changes in the thermospheric composition, winds and density, and its coupling to ion composition and density is critically dependent on real-time description of the these driving forces, the electric field, and the energy that it controls. To do this we must study IMF effects on convection electric fields low-latitude electric fields, and electric-field pattern evolution; study solar wind pressure effects on convection electric fields; study time-varying electric field patterns effects on neutral winds and propagation from dayside to nightside of polar cap; determine energy deposition real time variances and impacts on neutral density and composition; study mesoscale variations in electrodynamics and effects on global structure; define role of disturbance dynamo and fossil winds at middle and low latitudes; and study the Appleton anomaly.

Spacecraft:

Type: 38 microspacecraft (spinners)
1 EE class spacecraft (Dynamics Explorer-B type)

Special Features: 2 microspacecraft fly in tandem with EE-class spacecraft

Section 6 Candidate Future Missions

Special Requirements: Microsat design is volume constrained for launch operations
- Multiple spacecraft deployment mechanism
- Spin-up mechanism

Instruments:

	Mass (kg)	Power (W)	Data rate (bps)	Data storage	FOV
Micro S/C payload (38 S/C total)				24-hour storage of 500 bps data stream	
Vector Electric Fields	5.30	4.00	100		TBD
Vector Magnetometer	3.00	2.00	100		TBD
Low Energy Electron Detector	6.00	4.00	100		TBD
Micro S/C subtotal:	14.3	10.00	300		
"De-class" S/C payload (1 S/C)					
Vector Electric Field Instrument	7.00	8.00	100	TBD	TBD
Vector Magnetometer	3.00	4.00	100	TBD	TBD
Low Energy Ions/Electron	6.00	2.00	100		
Neutral Mass Spectrometer	8.00	12.00	300		
Neutral Winds & Temp Spect.	8.00	10.00	300	TBD	TBD
Ion Mass Spectrometer	6.00	8.00	300		
"De-class" S/C payload total	30.00	40.00	1200		

Mission Strategy:

An innovative approach is needed to provide global patterns with high temporal resolution (10 minutes). Mass production of satellites with a sounding rocket approach can provide multiple, low-cost measurements. Reproduction costs are estimated to be ~$2–3M per satellite. A dual approach is needed for this mission:

- Constellation of 36 "Micro-Sat" satellites to determine real time field gradient inputs (Class A measurement parameters)

- One "EE-class" satellite flown in tandem with two Micro-Sat satellites to correlate thermospheric and ionospheric measurements with field gradients (Class B measurement parameters).

This innovative approach means that the catastrophic loss of any one micro-sat does not seriously impact the mission. Hence lower reliability criteria can be used to minimize costs and maximize return. We anticipate that a number of the micro-sats can be launched from the same vehicle (possibly Pegasus), thus minimizing launch costs. The electric field maps would allow the investigation of thermosphere-ionosphere-magnetosphere coupling on a much more detailed level than will ever be possible using more conventional techniques (e.g., DE, DMSP, etc.). With simultaneous global auroral imaging, the subsequently inferred conductivities can be combined with the electric fields to produce a 2-D map or footprint of magnetospheric electrodynamics. Using techniques such as the Assimilative Mapping of Ionospheric Electrodynamics (AMIE)—a technique to merge observations of electric fields and currents in order to generate a self-consistent global model—the much needed dynamic magnetospheric inputs to the thermosphere and ionosphere would be obtained.

Technology Requirements: TBD

Points of Contact:

    Science Panel: Dr. Tim L. Killeen (313) 747-3435
Space Physics Research Laboratory
University of Michigan
2455 Hayward Street
Ann Arbor, MI 48109-2143

    Discipline: Ionospheric, thermospheric, & mesospheric physics

    Program Manager: TBD

    Program Scientist: Dr. Dave Evans (202) 453-1514

    NASA Center: JPL

    Project Manager: Ron Boain (818) 354-5122

    Project Scientist: TBD

6.16 Mercury Orbiter

Target: Mercury magnetosphere

Orbit: Mercury polar orbit, 200 km x 7 RM, multiple Venus & Mercury gravity assists

Mission Duration: 6 years

Mission Class: Moderate

Mass: 2 spacecraft @800 kg each

Launch Vehicle: Titan IV(SRMU)/Centaur

Theme: Magnetospheric physics

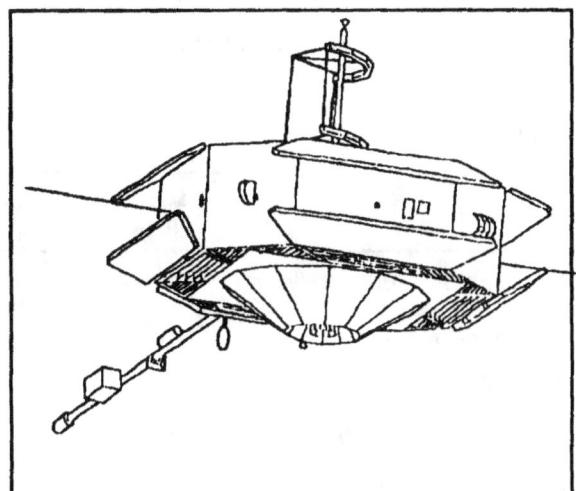

Science Objectives:

Magnetospheric physics

- Obtain a 3-D map of the miniature magnetosphere structure and plasma environment.
- Take advantage of the unique magnetospheric environment of Mercury (size and absence of appreciable atmosphere) to unravel the principal magnetospheric processes and to advance the understanding of the processes occurring at Earth.
- Investigate cause-effect relationships between the interplanetary conditions and the magnetospheric processes using two spacecraft.

Planetology:

- Complete imaging of Mercury at resolutions better than 1 km and more than 25% at resolutions better than 250 M.
- Obtain global geochemical (e.g., Fe, Th, K, Ti, Al, Mg and Si) terrain map.
- Measure detailed magnetic field and detect anomalies.
- Measure gravitational field.

Heliospheric physics:

- Study the inner heliospheric structure and dynamics.

Solar physics:

- Obtain X-ray, gamma-ray and neutron data of the Sun not obtainable at Earth.

Spacecraft:

Type: Spinners (2 spacecraft)
Special Features: One 7 m science boom; two 50 m wire booms
Special Requirements: N/A

Instruments:

	Mass (kg)	Power (W)	Data rate (kbps)	Data storage	FOV
DC Electric Field Analyzer	14.6	6	.064–8.5	TBD	TBD
Energetic Particle Detector	15	15	1–10	TBD	12° x 180° 50° x 180°
Fast Electron Analyzer	4	5	1–10	TBD	15° x 180°
Fast Ion Analyzer	4	5	1–10	TBD	15° x 180°
Gamma Ray Spectrometer	17	14.3	1.2–2.4	TBD	±10°/±20°
Ion Composition Plasma Analyzer	10	12	1–10	TBD	15° x 180°
Line Scan Imaging (& TEC)	5.1	11	10	TBD	0.015°x30°
Magnetometer	5.3	5.5	1–5	TBD	TBD
Optimized Solar Wind Analyzer	10	10	0.4–4	TBD	45° x 180°
Radio/Plasma Wave Analyzer	4.6	6.5	.032–10	TBD	TBD
Solar Neutron Analyzer	10	10	0.5	TBD	TBD

Mission Strategy:

Heliocentric Trajectory

- Multiple gravity-assists of Venus and Mercury
- Flight time: 3–4 yrs. for E-VV-MM-M or 4.5–6 Yrs.

Section 6 Candidate Future Missions

Mercury Phase Scenario

- Emphasize magnetosphere survey for the first two Mercury years.
- Emphasize imaging for the next two Mercury years.

Mercury Phase Orbits

S/C-1: • 200 km altitude X 12 hr. polar orbits. Periapsis at N-pole

S/C-2: • Initially very loose equatorial orbit for tail excursion
- Gradual orbit reduction for mid-tail survey
- Finally, orbit plane and size changed to 200 km x 12 hr. polar with periapsis at equator for two spacecraft imaging

Propulsion Requirements

C3 : 11–28 (km/s)^2 ΔV : 2.5–3.2 km/s (depend on launch year)

Launch Opportunities

Aug. 1997, Jul. 1999, Sep. 2002, Jul. 2004, Aug. 2005, Jul. 2007

Technology Requirements: None

Points of Contact:

Science Panel:	Dr. James A. Slavin (301) 286-5839 Code 696 Goddard Space Flight Center National Aeronautics and Space Administration Greenbelt, MD 20771
Discipline:	Magnetospheric physics
Program Manager:	TBD
Program Scientist:	Dr. Tom Armstrong (202) 453-1514
NASA Center:	JPL
Project Manager:	Ron Boain (818) 354-5122
Project Scientist:	TBD

Section 6 Candidate Future Missions

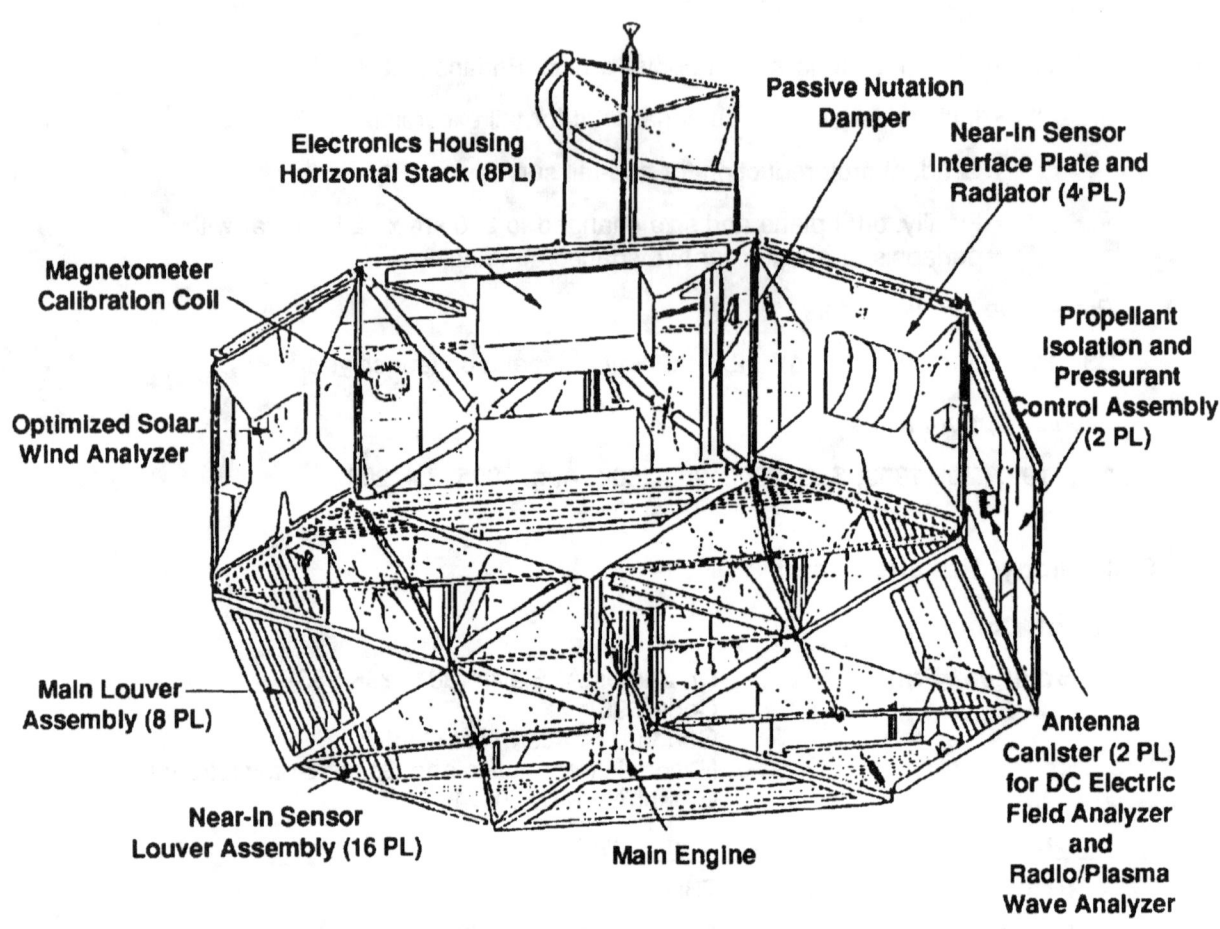

Mercury Orbiter spacecraft system internal configuration

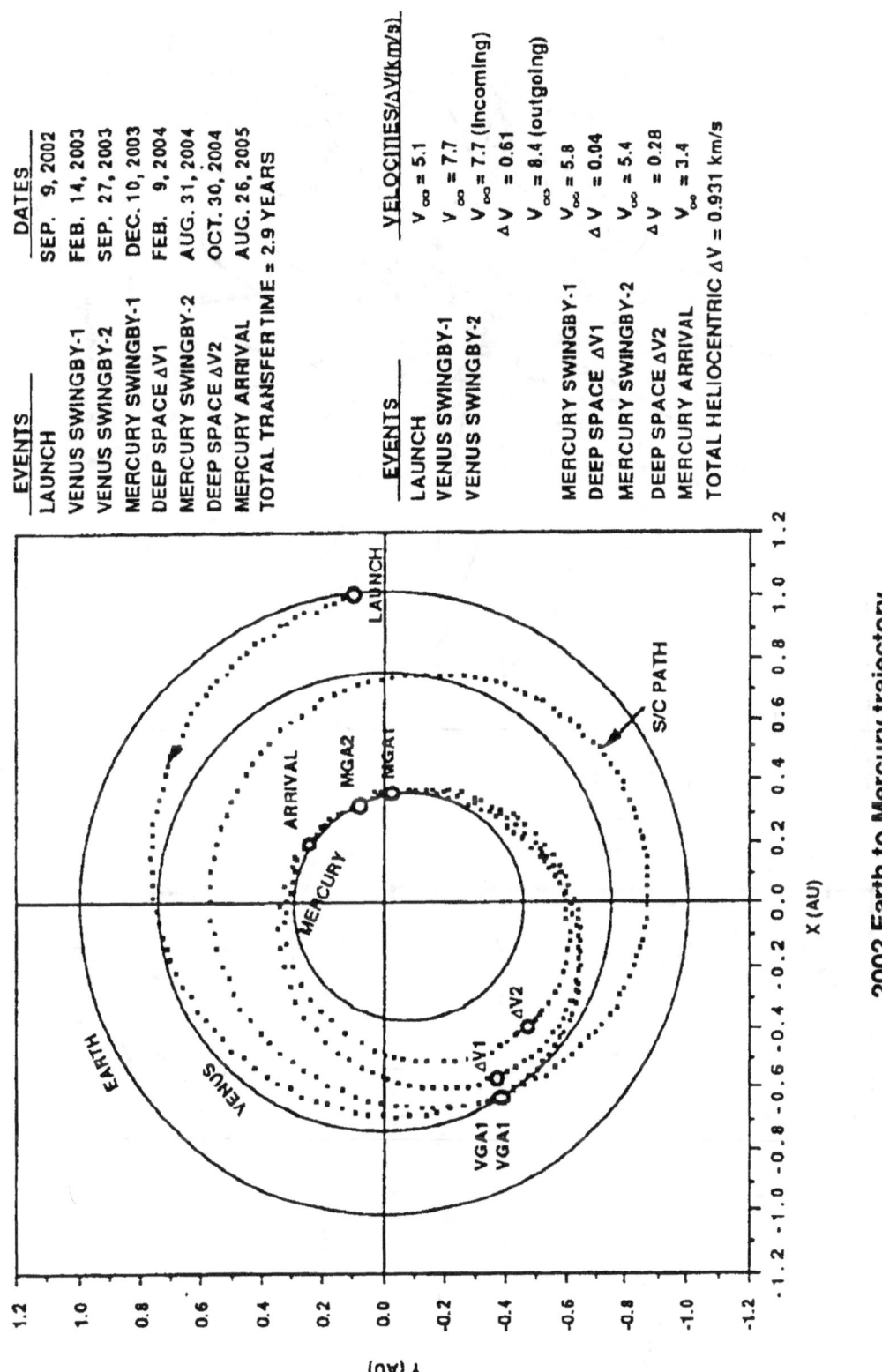

2002 Earth to Mercury trajectory (E-VV-MM-M)

Section 6 Candidate Future Missions

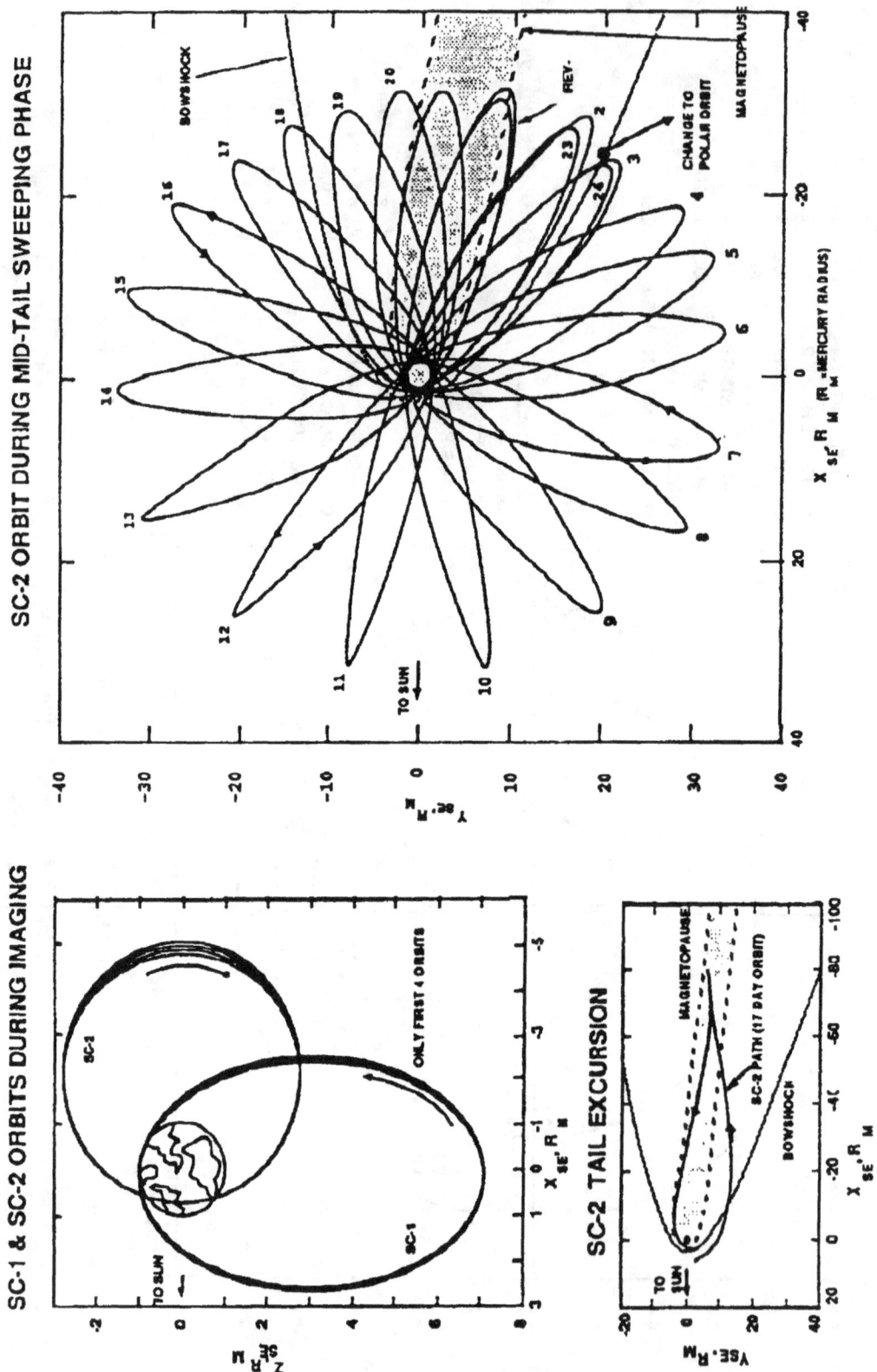

Mercury phase orbital design

6.17 Mars Aeronomy Observer

Target: Martian atmosphere

Orbit: Mars high inclination;
Highly elliptical
Periapsis: <150 km
Final orbit: 325 km x 6000 km, I = 63.3°

Mission Duration: Cruise: 0.8 years
Encounter: 1.9 years
Total: 2.7 years

Mission Class: Moderate

Mass: TBD

Launch Vehicle: Titan III/TOS

Theme: Comparative planetary atmospheres/ magnetospheres

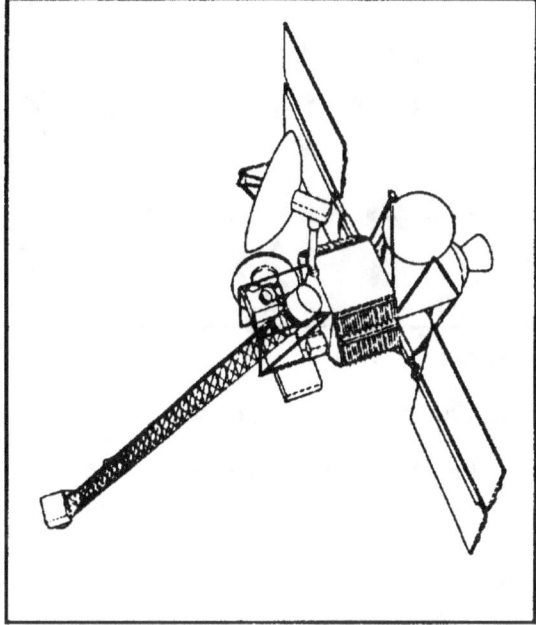

Science Objectives:

The goals of the MAO mission are to determine the diurnal and seasonal variations of the upper atmosphere and ionosphere; determine the nature of the solar wind interaction; study crustal magnetic and gravity anomalies; and measure the present thermal and non-thermal escape rates of atmospheric constituents and determine what these rates imply for the history and evolution of the Martian atmosphere. Specific objectives of the mission are to:

- Explore the thermosphere, ionosphere, and mesosphere of Mars and the effects of the interaction of the solar wind with these regions.

- Investigate the photochemistry, heat budget, and dynamics of the Martian upper atmosphere in the context of our understanding of such processes from similar missions already conducted at Earth and Venus.

- Examine the effects on the upper atmosphere of tropospherically generated atmospheric waves and dust storms.

- Test current aeronomic theory against the reality of a different atmosphere which is subject to the same physical processes, but in different combinations, and with different results (Comparative planetary study using our knowledge of Earth, Venus, Mars, Titan, etc.).

Section 6 Candidate Future Missions

Spacecraft:

    Type: 3-axis stabilized or spinner with despun platform
Sun-pointed, non-Sun synchronous

    Special Features: N/A

    Special Requirements: Sufficient ΔV to control periapsis altitude and accommodate plane changes and apoapsis decreases

Instruments:

	Mass (kg)	Power (W)	Data rate (bps)	Data storage	FOV
Baseline payload:					
Neutral Mass Spectrometer	10	8.5	180	TBD	TBD
Fabry-Perot Interferometer	13.5	5.5	30	TBD	TBD
UV & IR Spectrometer	5	7	130	TBD	TBD
Ion Mass Spectrometer	2.5	1.5	60	TBD	TBD
Retarding Potential Analyzer + Ion Driftmeter	4.5	4	80	TBD	TBD
Langmuir Probe	2	9	30	TBD	TBD
Plasma + Energetic Particle Analyzer	10	9	320	TBD	TBD
Magnetometer	3	3.5	200	TBD	TBD
Plasma Wave Analyzer	5.5	3.5	130	TBD	TBD
Radio science (radio occultation)	4.5	12.5	N/A	TBD	TBD
Optional Payload:					
Infrared Atmospheric Sounder	8	7.5	260	TBD	TBD
UV & Visual Synoptic Imager	9	8	1000	TBD	TBD
Neutral Winds & Temperature Spectrometer	10	9	180	TBD	TBD

Specific Measurements:
- Neutral, ion composition
- Neutral, ion, & electron temperatures & densities
- Neutral winds/ion drift speed
- Magnetic & electric fields & plasma wave environment
- Solar wind & energetic particles
- UV, optical, & IR limb scanning & imaging
- Radio science (radio occultation)

Mission Strategy:

The MAO mission will deliver a spacecraft to Mars for an extended orbital study of the planet's surface, atmosphere, and gravitational and magnetic fields. The spacecraft will operate in a polar elliptical orbit and collect aeronomy data on the magnetosphere, solar wind interactions, upper and lower atmosphere, and surface topography, and will then be transferred to a low (200–300 km) circular orbit. The eccentric phase of the mission employs a very low periapsis (110–150 km) to perform *in situ* and remote measurements to the lowest possible altitudes as well as measurements of the solar wind and the energetic plasma environment within the magnetosphere and upper ionosphere. The circular phase will provide measurements of the global response of the upper atmosphere over extended periods to examine its local time, seasonal, and solar cycle behavior.

Mission Profile:

Launch Opportunities:

Launch	Arrival	C3	MOI ΔV
11/15/96	09/14/97	9.08	0.960
12/14/98	10/01/99	10.47	1.229
02/01/01	10/26/01	13.26	1.522
06/08/03	12/29/03	8.82	0.876
08/06/05	07/18/06	17.95	0.952
09/12/07	08/20/08	13.55	0.785

Initial Orbit: 200 x 43,000 km, I = 102 degrees

Latitude Map Orbit: 150 x 6000 km, I = 102 degrees (MOI + 153 D)

Local Time Map Orbit: 150 x 6000 km, I = 63.3 degrees (MOI + 469 D)

Final Orbit: 325 x 6000 km, I = 63.3 degrees (MOI + 687 D)

Section 6 Candidate Future Missions

Technology Requirements: None

Points of Contact:

    Science Panel: Dr. James A. Slavin (301) 286-5839
                                    Code 696
                                    Goddard Space Flight Center
                                    National Aeronautics and Space Administration
                                    Greenbelt, MD 20771
                                    Dr. Edward P. Szuszczewicz (703) 734-5516
                                    (ITM Panel point of contact)

    Discipline: Aeronomy/magnetospheric physics
    Program Manager: TBD
    Program Scientist: Dr. Tom Armstrong (202) 453-1514
    NASA Center: JPL
    Project Manager: Ron Boain (818) 354-5122
    Project Scientist: TBD

Section 6 Candidate Future Missions

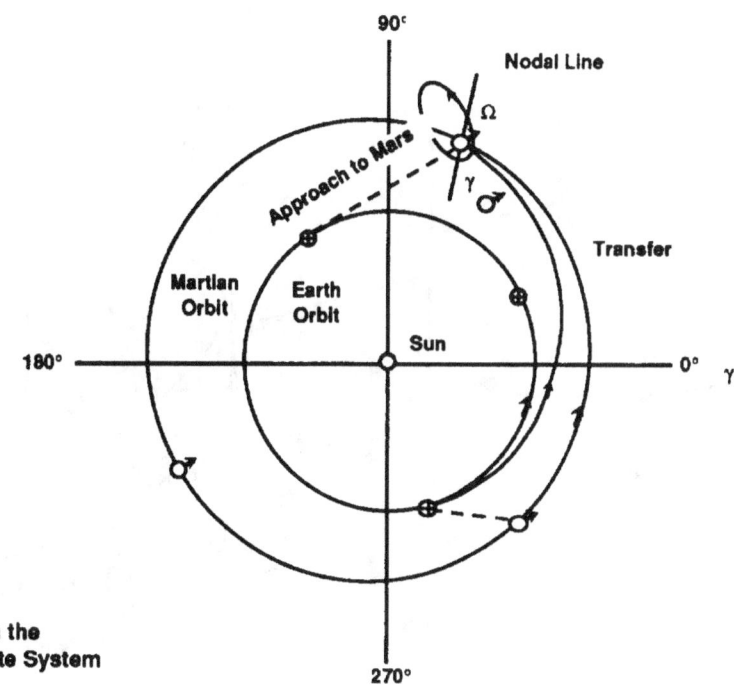

Orbits of Earth and Mars:
Position of Earth and Mars at the spacecraft launch instant ("Start") and at the time of the spacecraft approach to mars ("Approach to Mars");

Earth-to-Mars transfer trajectory first orbit of the spacecraft around Mars;

Nodal line of the spacecraft's first orbit

Notations:
- ⊕ Earth
- ○ Mars
- γ Vernal Equinox Point in the Areocentric Coordinate System
- Ω Ascending Node

Mars Orbiter transfer orbit

Section 6 Candidate Future Missions

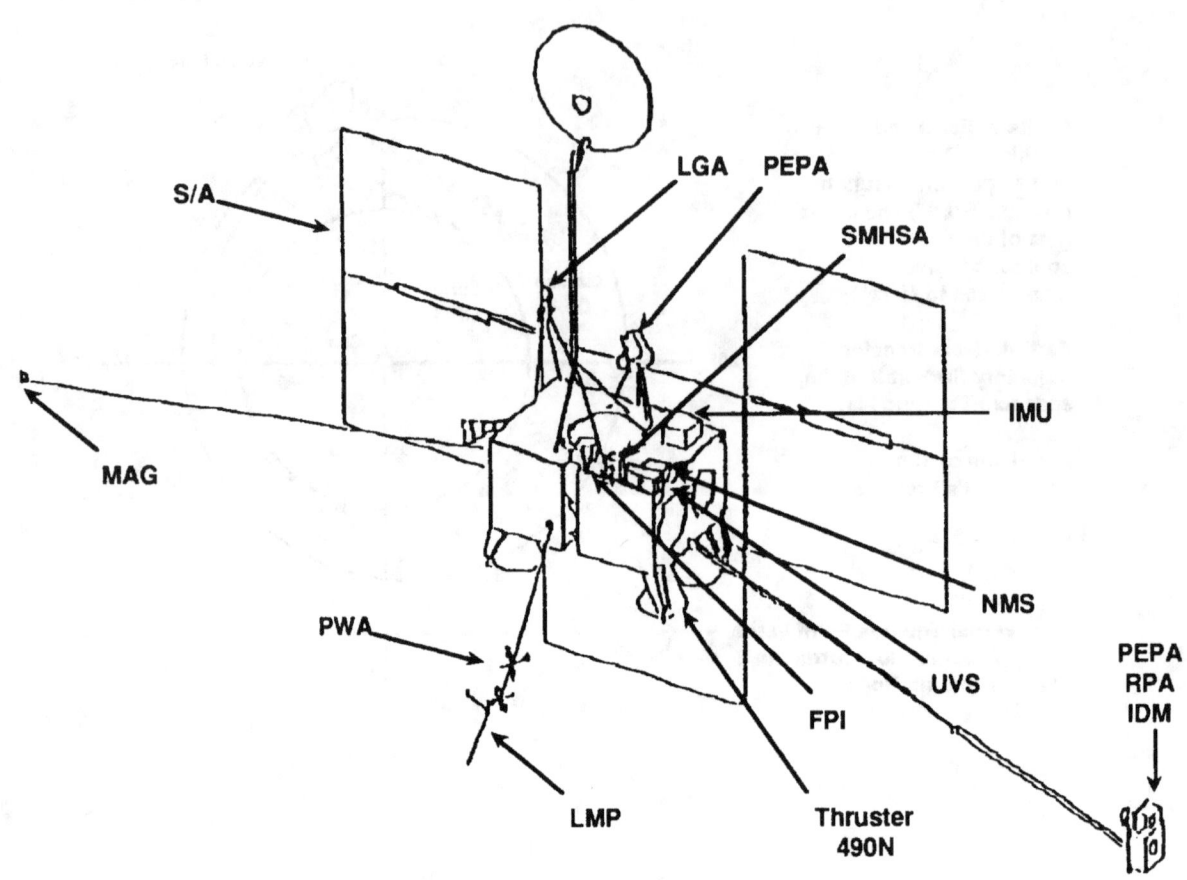

HGA	High Gain Antenna
LGA	Low Gain Antenna
S/A	Solar Array
CSA	Celestial Sensor Assembly
SMHSA	Edge Scanner, Mars Horizon Sensor
IMU	Inertial Measurement Unit

**Mars Aeronomy Orbiter
mission configuration concept**

6.18 Global Solar Mission (Polar Orbiter & 1 AU In-Ecliptic Network)

Target: The Sun; 1 AU particles & fields

Orbit: Solar polar & in-ecliptic, circular @1 AU
Polar inclination: ≥70°
In-ecliptic inclination: ~0°
(one in-ecliptic spacecraft is in LEO)

Mission Duration: Polar: 2–3 years cruise
3+ years operation
In-ecliptic: TBD

Mission Class: Major

Mass: Earth Orbiter: 2400 kg
Solar Orbiter: 5850 kg

Launch Vehicle: Earth Orbiter: Delta
Solar Orbiter: Titan IV/Centaur

Theme: Global view of solar magnetic activity and influence on the solar system; determination of the mechanisms and processes governing the magnetic activity cycle of the Sun

Science Objectives:

Solar Polar Orbiter scientific objectives are to observe the global Sun; measure polar magnetic fields accurately; measure high latitude differential rotation; investigate pole-equator temperature differences; observe coronal streamers and mass ejections from above; measure particles and fields in the solar wind as a function of latitude; and measure helioseismology p-modes free of rotational splitting from the pole.

In-Ecliptic Network scientific objectives are to study the magnetic activity cycle and evolution of the solar field at a variety of spatial and thermal regions; and study the role of the Sun in shaping the structure and dynamics of the heliosphere.

Spacecraft:

Type: Polar: 3-axis stabilized or spinner with despun platform (1 spacecraft)
In-Ecliptic: 3-axis stabilized (2 spacecraft)

Special Features: Polar: Pointing to Sun center within 10 arc sec; pointing stability at least to 3 arc sec r.m.s.
In-Ecliptic: Pointing to Sun center within 10 arc sec; pointing stability at least to 0.3 arc sec r.m.s.

Special Requirements: Spacecraft and instrument cleanliness (volatiles and particulates) must be maintained.

System Design Concept: The IEN spacecraft are basically "Solar Max's" with a new science payload. That is, they are 3-axis stabilized vehicles using the (relatively heavy) modular MMS construction. The mass of the Earth orbiter is approximately 2400 kg, while the solar orbiter has an additional 3450 kg for the large propulsion system to perform the circularization maneuver.

Subsystem Descriptions:

Attitude Control Subsystem: Two 2-axis fixed-head star trackers (8° x 8° FOV) and three 2-axis gyros (0.02 arc sec resolution). Reaction wheels used to control attitude. Solar Max met the IEM requirement of 10 arc sec point accuracy and 0.3 arc sec rms stability. The solar orbiter requires a small attitude control subsystem for desaturation of the reaction wheels, since the magnetic torquer designed for use in Earth orbit will not function away from Earth's magnetic field.

Power Subsystem: Two fixed solar paddles supplying 1500-3000 W, with three Ni-Cad batteries, 20 A-hr each. Whether the structural support for the solar paddles will have to be beefed up for the circularization maneuver is TBD.

Propulsion Subsystem: The solar orbiter requires a large bipropellant propulsion module for the orbit circularization maneuver of 2 km/s. The propellant mass is 2925 kg, plus 459 kg for tanks and 66 kg for the motor/nozzles/structure. The assumption that the MMS structure will be able to support the loads during the maneuver will need to be verified by analysis. There should be little penalty in using a fairly low thrust level, since there is nothing critical about the timing of the maneuver.

Structure and Mechanisms Subsystem: Structure is the modular Multi-Mission Spacecraft type. The observatory is approximately 4 meters long (plus the solid rocket module), and fits into a circular envelope 2.5 meters in diameter. The top 2.3 meters comprises the instrument module, including the Fine-Pointing Sun Sensor.

Command and Data Handling Subsystem: Data handling is the only area where Solar Max falls short, but there is does so grandly. It could handle a maximum downlink rate of 512 kb/s, which is 1/20 the rate of IEM's smaller instruments. Data compression techniques will probably be needed for the Earth orbiter, and definitely will be needed for the solar orbiter.

Instruments:

	Mass (kg)	Power (W)	Data rate	Data storage	FOV
Polar instruments:					
X-Ray Imager	20	15	20 kbps		
WL Coronograph	20	15	20	TBD	TBD
Magnetograph/Doppler Telescope	30	20	40	TBD	TBD
Solar Irradiance	10	7	1	TBD	TBD
Zodical Light Photometer	5	5	1	TBD	TBD
Particles & Fields	TBD	TBD	TBD	TBD	TBD
In-Ecliptic instruments:					
WL Coronograph	50	50	10 mbps	TBD	TBD
Vector Magnetograph/ Dopplergraph	100	100	20 mbps	TBD	TBD
UV Coronograph/Telescope with Spectrograph	100	100	20 mbps	TBD	TBD
Helioseismology Instrument	50	50	10 mbps	TBD	TBD

White-Light Coronograph (WLC):

Measurement Objectives: The 3-D structure of the solar corona in the region 0.5 to 6 solar radii above the photosphere.

Principle of Operation: In a normal telescope imaging the inner corona would be impossible because of light from the photosphere scattered in the instrument would obscure the corona. The WLC uses an external occulting disk to keep much of the photospheric light out of the instrument, and internal baffles, occulting disks, and lens coatings to handle that which does get in.

Accommodation and Integration Requirements: The WLC requires mounting on the spacecraft with a clear field of view. It is 2.5 meters long, and 25 cm in diameter.

Technology Readiness: The telescope and sensor (a CCD with 2000x2000 pixels) use conventional technology and require no new developments. The spacecraft in solar orbit will need image processing capabilities in order to reduce the data rate. The amount of technology development needed to shoehorn the data into a reasonable link is TBD.

Design Heritage: Similar instruments are on the Solar Maximum Mission and SOHO.

Vector Magnetograph/Dopplergraph (VM/D):

Measurement Objectives: The vector magnetic field at the surface of the Sun, and flow of near-surface materials at the scales of microturbulence and macroturbulence.

Principle of Operation: The wavelengths of spectral lines are shifted by motion along the line of sight to an observer, and they may be split by magnetic fields (some are, and some aren't—it depends on the transition which generates the line). The VM/D forms images of the Sun at two wavelengths very close to each other, and uses the difference image to determine either the magnetic field or the motion. The VM/D is a telescope using either filters or a spectrograph to separate the wavelengths.

Accommodation and Integration Requirements: The VM/D requires mounting on the spacecraft with a clear field of view. It is about 2.5 meters long and 80 cm in diameter. The telescope aperture is 30-60 cm.

Technology Readiness: The spacecraft in solar orbit will need image-processing capabilities in order to reduce the data rate. The amount of technology development needed to shoehorn the data into a reasonable link is TBD.

Design Heritage: The VM/D is similar to the Solar Optical/UV Polarimeter (SOUP) instrument to be flown on Spacelab II, but the SOUP mirror is too small.

UV Coronograph/Telescope with Spectrograph (UVCT):

Measurement Objectives: High-temperature phenomena in the corona and against the disk of the Sun.

Principle of Operation: A concave mirror with a concave grating is used to form the image.

Accommodation and Integration Requirements: The UVCT requires mounting on the spacecraft with a clear field of view. The UVCT is 2–2.5 meters long, with a cross section approximately 40x70 cm.

Technology Readiness: There are several items which need to be developed: near-normal-incidence optics, toroidal holographic gratings, and detectors which reach to the extreme UV with the needed sensitivity and small-enough pixels. The spacecraft in solar orbit will need image processing capabilities in order to reduce the data rate. The amount of technology development needed to shoehorn the data into a reasonable link is TBD.

Design Heritage: Some similarity to the S055 experiment on Skylab's Apollo Telescope Mount.

Helioseismology Instrument (HSI):

Measurement Objectives: Solar oscillations (helioseismicity) without the ambiguities imposed by having only a single viewing location.

Principle of Operation: Use Doppler techniques to measure radial motion of solar material. The HSI is similar to the VM/D.

Accommodation and Integration Requirements: The HSI requires mounting on the spacecraft with a clear field of view. It is 1.5 meters long and 40 cm in diameter.

Technology Readiness: The spacecraft in solar orbit will need image processing capabilities in order to reduce the data rate. The amount of technology development needed to shoehorn the data into a reasonable link is TBD.

Design Heritage: Similar to the MDI on SOHO.

Mission Strategy:

The In-Ecliptic Network consists of one spacecraft in Earth orbit and one in the same heliocentric orbit as Earth, but trailing by 90° so that it sees the side of the Sun which is rotating toward Earth. Two more solar orbiters may be added (one each at the 180° and 270° points) as solar flare warning detectors in support of SEI. If the additional spacecraft are not flown, it might be desirable to put the solar orbiter somewhat farther from Earth, so that its view of the Sun has less overlap with the Earth's orbiter's. The Solar Polar Orbiter uses advanced propulsion to achieve a polar solar orbit at 1 AU. *In-situ* fields and instruments operate through entire mission to explore the latitudinal structure of the heliospheric fields and flows through the mission lifetime. The zodical light instrument uses low spatial resolution photometry to detect variations in the distribution of material located in the solar system. The polar orbit will allow long periods during which solar structures may be observed constantly so that evolution is not modulated by solar rotation.

Technology Requirements:

Data compression (or alternatively, data storage and transmission) is the only recognized problem. The needs (bringing back 40 Mb/s from 1.4 AU) seem to go well beyond what can be done noiselessly with current techniques. The algorithms which are being developed and implemented in silicon need to be evaluated for applicability to these data. When the required telemetry rate for data with an acceptable noise level has been determined, then tradeoffs between cost and data quality/quantity can be made.

Points of Contact:

Science Panel:	Dr. Richard R. Fisher (303) 497-1566 High Altitude Observatory National Center for Atmospheric Research P.O. Box 3000 Boulder, CO 80307
Discipline:	Solar physics
Program Manager:	TBD
Program Scientist:	Dr. Dave Bohlin (202) 453-1514
NASA Center:	JPL
Project Manager:	Ron Boain (818) 354-5122
Project Scientist:	TBD

Section 6 Candidate Future Missions

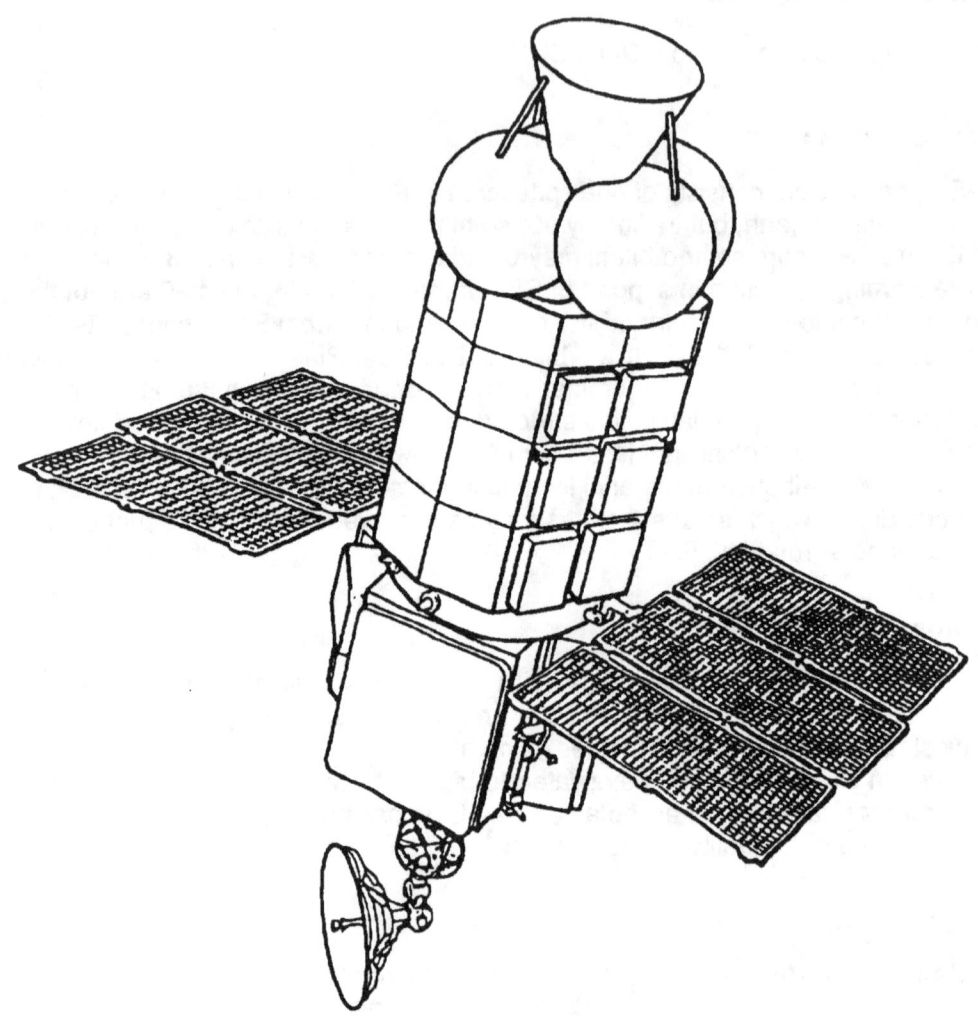

**In-Ecliptic Network – Solar Orbiter
pre-circularization configuration**

Section 6 Candidate Future Missions

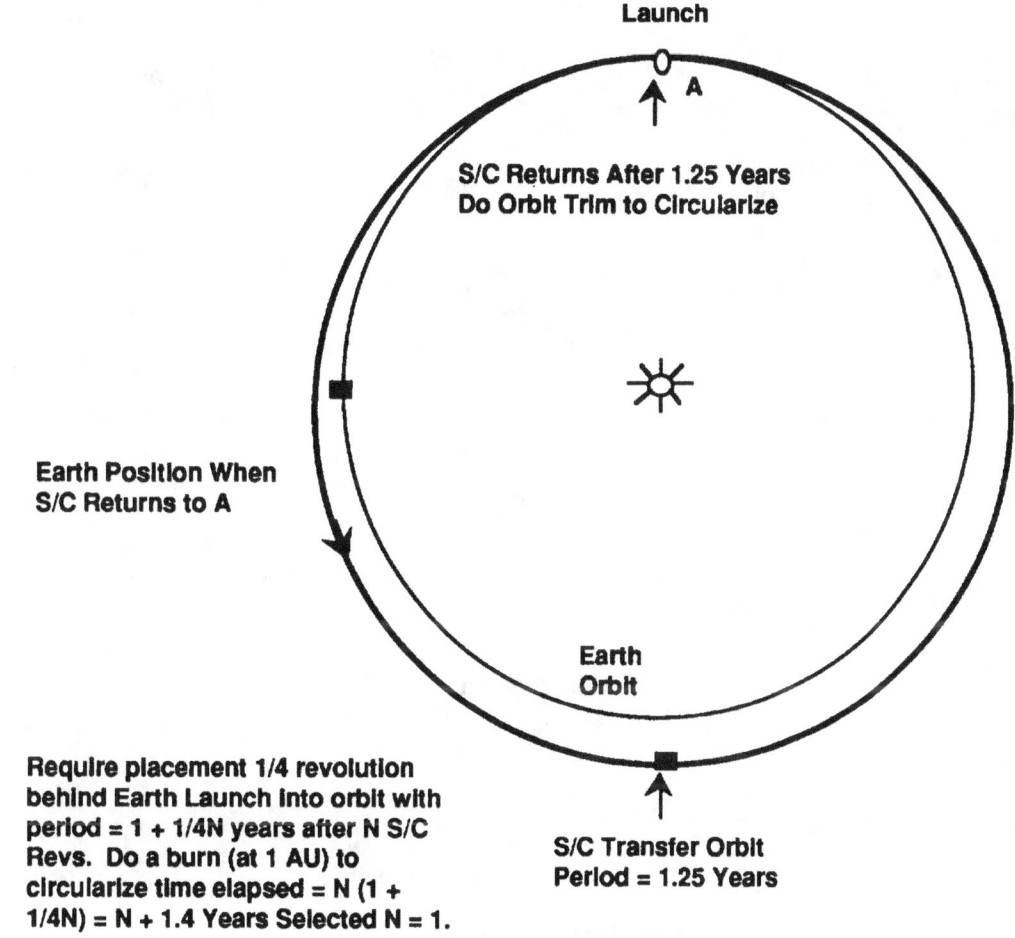

In-ecliptic network trajectory

Section 6 Candidate Future Missions

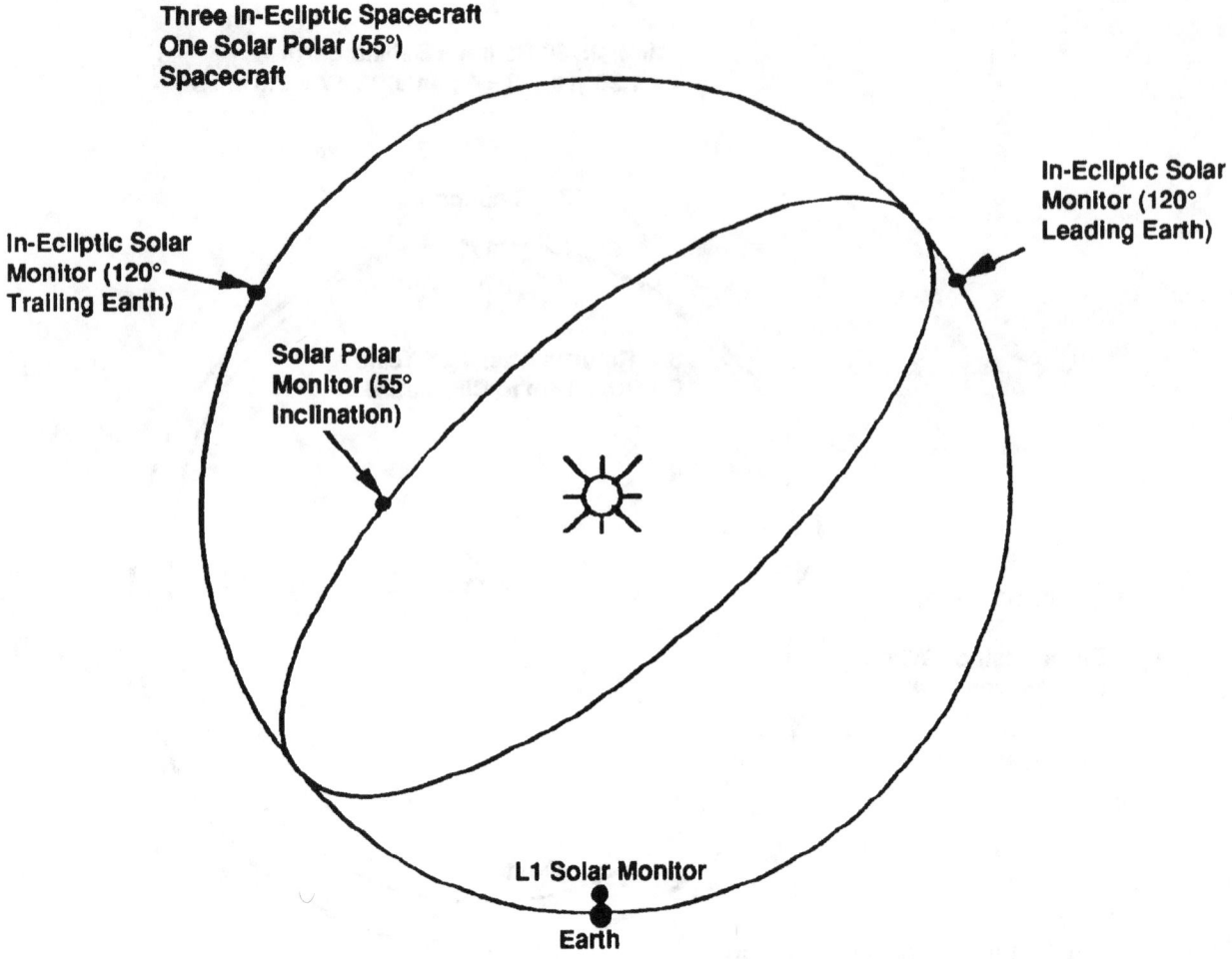

Global Solar Mission (GSM)

Appendix A
Launch Sites

Appendix A Launch Sites

Launch Sites

Launch Base	Country	Location Latitude	Longitude	Launch Azimuth (Approx. Range)	Launch Pads
Kennedy Space Ctr/CCAFS	USA	28° 28' N	80° 33' W	35-120°	SLC-17A (Delta II) SLC-17B (Delta II) SLC-36A (Atlas II) SLC-36B (Atlas I/Atlas II) SLC-37 (EPAC S-Series Activation) LC-39A (Shuttle) LC-39B (Shuttle) SLC-40 (Titan III/Titan IV) SLC-41 (Titan IV)
Vandenberg AFB	USA	34° 35' N	120° 38' W	140-208°	SLC-2W (Delta, Inactive) SLC-3E (Atlas, Inactive) SLC-3W (Atlas E) SLC-4E (Titan 34D/Titan IV) SLC-4W (Titan II) SLC-5 (Scout) SLC-6 (Shuttle, Inactive) *SLC-X (Titan IV)
Wallops Island	USA	37° 52' N	75° 27' W	90-109°/126-129°	Launch Area No. 3 (Scout) (*Conestoga)
Dryden FRC/Edwards AFB	USA			0-360°	(Pegasus on NB-52B)
San Marco	Kenya (Italy)	2° 54' N	40° 20' E	82-130°	San Marco Platform (Scout, Scout II, *Conestoga)

* Proposed   ** Data Not Currently Available

Appendix A Launch Sites

Launch Sites

Launch Base	Country	Location Lattitude	Longitude	Launch Azimuth (Approx. Range)	Launch Pads
Kourou (Guiana Space Ctr)	Fr. Guiana (ESA)	5° 12' N	52° 39' W	350-108°	ELA-1 (AR 3, AR 4 - Back-up Only) ELA-2 (AR 4) *ELA-3 (AR 5)
Tanegashima	Japan	30° 23' N	130° 58' E	30-165°	(H-I, H-II)
Kagoshima (Uchinoura)	Japan	31° 15' N	131° 04' E	**	(M-3SII)
Baikonur (Tyuratam)	USSR	45° 55' N	63° 21' E	0-120°	(Proton D-1, Proton D-1e, Tsyklon F-1, Energia-1 Pad, Energia/Buran-2 pads, Soyuz A-2)
Plesetsk	USSR	62° 54' N	40° 39' E	**	(Tsyklon F-2, Soyuz A-2, Kosmos C-1, Molniya A-2e, **)
Kapustin Yar	USSR	48° 20' N	45° 51' E	**	(Kosmos C-1)
Sriharikota	India	13° 44' N	80° 14' E	**	(SLV-3, ASLV, PSLV, GSLV)
Shuang ch'eng Tzu (Jiquan)	China	40° 25' N	99° 44' E	**	(CZ-1D, CZ-2E)
Xi-Chang	China	27° 53' N	102° 16' E	90-220°	(CZ-3, CZ-3A)
Tai Yuan	China	(270 mi. SW of Biejing)		**	(CZ-4)
Negev	Israel	**	**	**	(Shavit)
*Cape York	Australia	~12° S	~143° E	**	(Zenit J-1e, **)
*Alcantara	Brazil	2° 20' S	44° 29' W	**	(VLS)

* Proposed   ** Data Not Currently Available

Appendix A Launch Sites

Sounding rocket launch sites

Appendix B
Launch Vehicle Performance

Appendix B Launch Vehicle Performance

Launch Vehicle Performance

Launch Vehicle	Sponsor/ Contractor	Initial Launch Capability	Launch Costs	Fairing Diameter	Maximum Payload (Kgs)			
					LEO (100nmi) I=minimum	I=90	Synch Alt Transfer	Earth Escape (Parabolic)
United States:								
Scout G-1 *	Ling-Temco-Vought	1989	$13-17M	0.86-1.06m	261	198	73	-
Liberty 1A	Pacific American	Funding + 21mos.	$6-7M	2.44m	222	172	-	-
Liberty 1B	Pacific American	Funding + 21mos.	**	2.44m	998	694	**	**
ILV-S	American Rocket Co	N/A	$9.5-10.5M	1.22m	340	272	**	**
ILV-I	American Rocket Co	N/A	$15-17M	2.29m	1,814	1,360	**	**
Pegasus	Orbital Sciences Corp	1989	$7.5-8.5M	1.27m	431	318	125	**
Taurus	Orbital Sciences Corp	1991	$20-23M	1.47-2.03m	1,678	1,361	204	**
Conestoga II	Space Services Inc	Funding + 14mos.	$13-28M	1.45m	635	454	-	**
Conestoga III	Space Services Inc	Funding + 14mos.	$13-28M	1.45m	998	**	-	**
Conestoga IV	Space Services Inc	Funding + 14mos.	$13-28M	1.45m	1,450	998	-	**
Conestoga V	Space Services Inc	Funding + 14mos.	$13-28M	1.45m	2,495	**	1,225	**
Dorado	CEROS Aerospace Corp	1992	**	**	1,585	**	**	**
S-I	E-Prime Aerospace Co	1991	$26-29M	2.32m	1,136	726	443	307
S-II	E-Prime Aerospace Co	1992	$38-42.5M	2.32m	3,000	**	1,071	739
S-II/GTK	E-Prime Aerospace Co	**	$44.5-50M	2.32m	3,500	**	1,250	861
S-III	E-Prime Aerospace Co	1993	$58.5-65.5M	3.05m	7,210	**	2,636	1,318
S-IV	E-Prime Aerospace Co	1993	$67.5-75M	**	9,240	**	3,568	2,045
S-V	E-Prime Aerospace Co	Under Study	$77.5-86.5M	**	11,200	**	4,400	2,585
S-VI	E-Prime Aerospace Co	Under Study	$94-105M	**	13,600	**	5,250	3,693
SEALAR	U.S. Navy/Truax Engr.	Under Study	$5M+	**	4,535	**	**	**

** Data Not Currently Available   * Scout II listed in European Section

Appendix B Launch Vehicle Performance

Launch Vehicle Performance

Launch Vehicle	Sponsor/ Contractor	Initial Launch Capability	Launch Costs	Fairing Diameter	Maximum Payload (Kgs) LEO (100nmi) I=minimum	I=90	Synch Alt Transfer	Earth Escape (Parabolic)
United States:								
Thor (w/upper stage)	USAF/McDonnell Douglas	9 in storage	**	1.65m		296		
Delta II (7920)	McDonnell Douglas	1990	$51-71M	2.90m	5,039	3,819	1,270	
Delta II (7925)	Mc Donnell Douglas	1990	$51-71M	2.90m			1,819	1,270
Atlas E	USAF/General Dynamics	(Thru 12/93)	$29-36M	2.13m	1,089	794		
w/upper stage	USAF/General Dynamics	(Thru 12/93)	$34-46M	2.13m		1,724	454	295
Atlas I	General Dynamics	1989	$71-80M	3.30-4.19m	5,900		2,340	1,520
w/upper stage	General Dynamics	1989	$76-90M	3.30-4.19m				**
Atlas II	General Dynamics	1991	$83-92M	3.30-4.19m	6,780		2,770	1,940
w/upper stage	General Dynamics	1991	$88-102M	3.30-4.19m				**
Atlas IIA	General Dynamics	1992	$89-99M	3.30-4.19m	7,120		2,900	2,100
w/upper stage	General Dynamics	1992	$94-109M	3.30-4.19m				**
Atlas IIAS	General Dynamics	1992/93	$108-121M	3.30-4.19m	8,610		3,630	2,870
w/upper stage	General Dynamics	1992/93	$113-131M	3.30-4.19m				**
Titan II	USAF/Martin Marietta	1988	$42-78M	3.05m	2,200	1,905		**
w/upper stage	USAF/Martin Marietta	1988	$47-88M	3.05m	3,350	2,850		**
Titan II (min refurb)	USAF/Martin Marietta	1990	$19-28.5M	2.44-3.05m	2,200	1,905		**
Titan II/MSX	USAF/Martin Marietta	1992	$39M+	3.05m	**	3,400		**
Titan III	Martin Marietta	1989	$165-185M	3.05-3.65m	14,742	12,519		
w/OIS (III-T)	Martin Marietta	1990	**	3.05-3.65m		**	4,990	3,630
Titan III (SRMU)	Martin Marietta	1991	$165-185M	3.05-3.65m	17,237	**		
Titan IV	USAF/Martin Marietta	1989	$230-276M	5.08m	17,690	14,515		4,627
w/Centaur	USAF/Martin Marietta	1990	$280-326M	5.08m	**	**		**
Titan IV (SRMU)	USAF/Martin Marietta	1991	$230-276M	5.08m	21,319	17,509		6,124
w/Centaur	USAF/Martin Marietta	1991	$280-326M	5.08m	**	**		**
Shuttle	NASA/Rockwell Int'l	1988	$300-345M	4.57m	24,721			**
Shuttle C	NASA/Rockwell Int'l	Under Study	**	4.57-7.62m	45K-77K			**

** Data Not Currently Available

Appendix B Launch Vehicle Performance

Launch Vehicle Performance

Launch Vehicle	Sponsor/ Contractor	Initial Launch Capability	Launch Costs	Fairing Diameter	Maximum Payload (Kgs)			
					LEO (100nmi) i=minimum	i=90	Synch Alt Transfer	Earth Escape (Parabolic)
United States:								
ALS	USAF/7 contractor study	Under Study	**	**	Heavy Lift	**	**	**
Single Stage to Orbit	SDIO (RFP Released)	**	**	**	**	**	**	**
Europe:								
Mariane	Sweden/France	Early 1990s	**	**	**	1,800	**	**
Scout II	ASI/Ling-Temco-Vought	1993	$19-21M	1.35m	536	419	122	31
AR 40	ESA/Arianespace	1989	$127-156M	4.00m	4,600	2,700	1,900	**
AR 42P	ESA/Arianespace	1989	**	4.00m	5,000	3,400	2,600	**
AR 44P	ESA/Arianespace	1989	**	4.00m	**	4,100	3,000	**
AR 42L	ESA/Arianespace	1989	**	4.00m	**	4,500	3,200	**
AR 44LP	ESA/Arianespace	1989	**	4.00m	**	5,000	3,700	**
AR 44L	ESA/Arianespace	1989	**	4.00m	9,550	7,500	4,200	2,200
AR 50	ESA/Arianespace	1995	**	5.00m	20,000	**	6,800	4,000
Japan:								
H-I	Japan/NASDA	1987	**	**	**	**	**	**
H-II	Japan/NASDA	1993	$127-170M	4.00-4.60m	9,997	**	3,996	1,996
M-3SII	Japan/ISAS	1987	**	**	815@200nmi	**	**	**
India:								
SLV-3	India	1980	**	**	35@200nmi	**	**	**
ASLV	India	1988	**	**	150@200nmi	**	**	**
PSLV	India	1990	**	**	**	998@500nmi	**	**
GSLV	India	1994	**	**	**	**	1,995	**

** Data Not Currently Available

Appendix B Launch Vehicle Performance

Launch Vehicle Performance

Launch Vehicle	Sponsor/ Contractor	Initial Launch Capability	Launch Costs	Fairing Diameter	Maximum Payload (Kgs) LEO (100nmi) i=minimum	i=90	Synch Alt Transfer	Earth Escape (Parabolic)
Soviet Union:								
Start	USSR	1991	$5-7M	**	300	**	**	**
Kosmos C-1 (SL-8)	USSR	1964	N/A	**	1,350	**	**	**
Tsyklon F-1 (SL-11)	USSR	**	**	**	4,000	**	**	**
Tsyklon F-2 (SL-14)	USSR	1977	N/A	**	5,500	**	**	**
Vostok A-1 (SL-3)	USSR	1959	N/A	**	4,730	**	**	**
Molniya A-2e (SL-6)	USSR	**	N/A	**	**	**	1600@SEO	**
Soyuz A-2 (SL-4)	USSR	**	**	**	**	**	7240@SEO	4,200
Zenit J-1 (SL-16)	USSR	1985	$81M+	**	13,740	**	3,800	**
Zenit J-1e	USSR	1992	**	**	15,700@12°	**	5,860	**
Proton D-1 (SL-13)	USSR	**	$38-92.5M	3.66m	22,680	**	10,000	2,000
Proton D-1e (SL-12)	USSR	**	$38-92.5M	3.66m		**	**	5,700
Energia K-1 (SL-17)	USSR	1987	**	5.5m	105,000	**	22,600	32,000
Energia/Buran K-1	USSR	1988	**	**	30,000	**	**	**
Energia (SL-W-2)	USSR	**	**	5.5m	160,000	**	**	**
China:								
Long March CZ-1D	China	1991	**	**	700-750	**	**	**
Long March CZ-2C	China	1982	**	**	2,500	**	**	**
Long March CZ-3	China	1984	$30M+	2.60m	3,289	**	1,400	**
Long March CZ-3A	China	1992	**	**		**	2,500	**
Long March CZ-2E	China	1990	**	2.20-4.0m	8,800	**	2,948	**
Long March CZ-4A	China	1988	**	**	4000@200nmi	2500@SSO	**	**
VLS	China/Brazil	Early 1990s	**	**	160@410nmi	**	**	**
Israel:								
Shavit	Israel	1988	**	**	170+	**	**	**

** Data Not Currently Available

Appendix B Launch Vehicle Performance

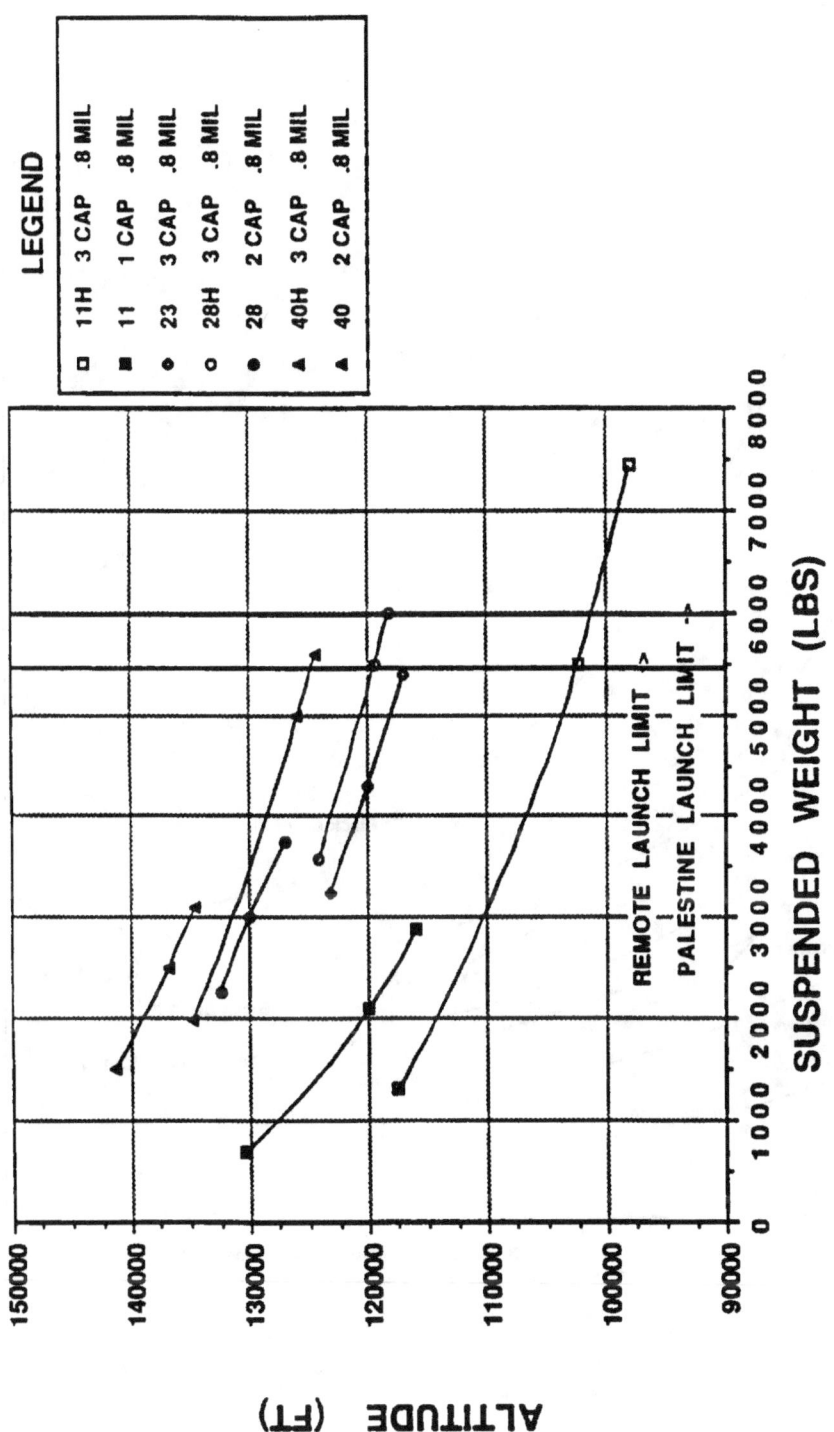

NASA balloon performance load-altitude curves

B-5

Appendix B Launch Vehicle Performance

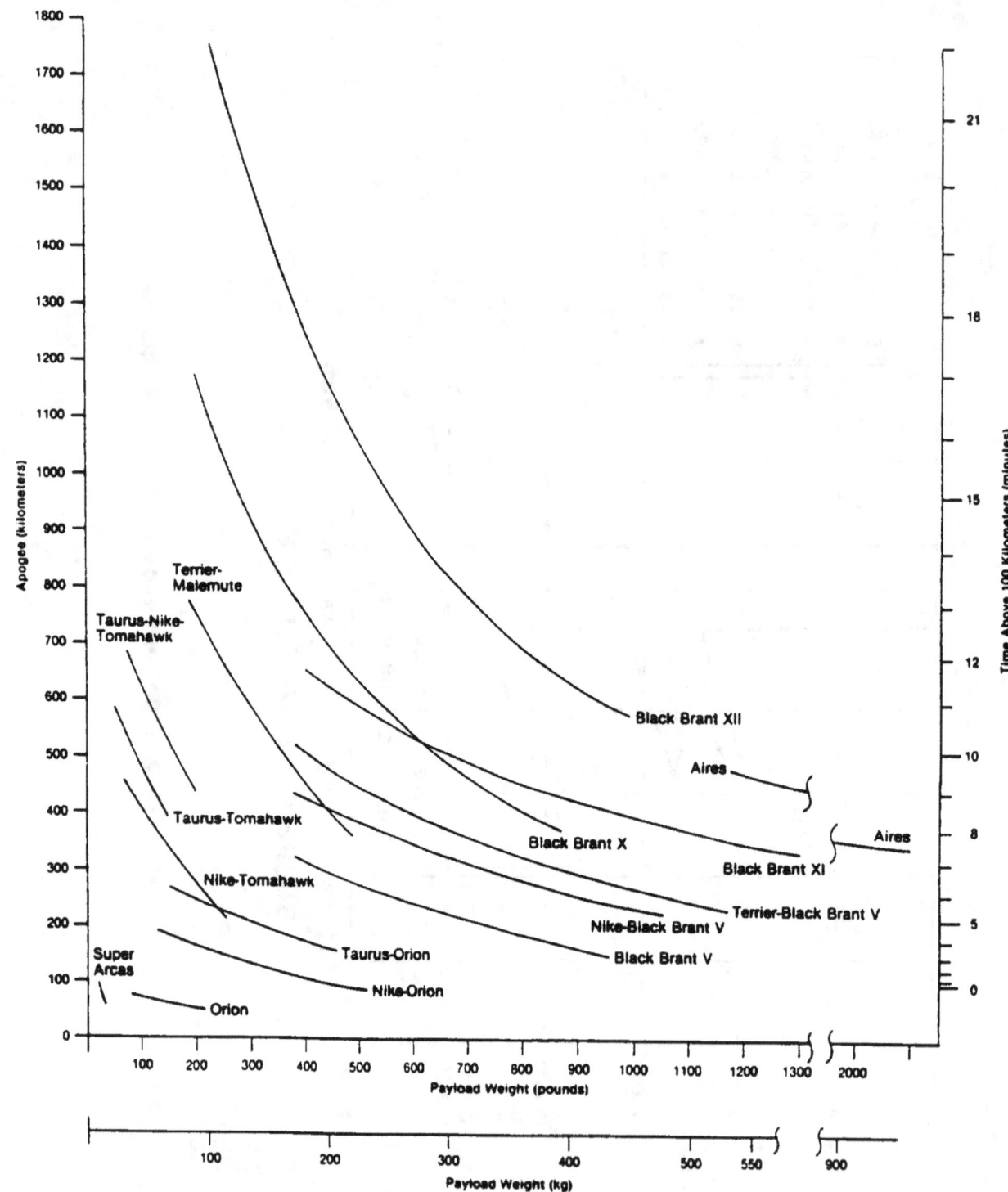

NASA sounding rocket performance

B-6

Appendix C
Acronyms

Acronyms and Abbreviations

ABM	Apogee Boost Motor
ACE	Advanced Composition Explorer
	Attitude Control Electronics
ACT	Attitude Control Thrusters
ACTIVE/APEX	A Soviet Multinational Plasma Wave Experiment
AMPTE	Active Magnetospheric Particle Tracer Explorers
AE	Atmosphere Explorer
AO	Announcement of Opportunity
AOCS	Attitude and Orbital Control System
AU	Astronomical Unit
AURA	Association of Universities for Research in Astronomy
BER	Bit Error Rate
bps	Bits Per Second
CCE	Charge Composition Explorer
CDR	Critical Design Review
CIB	Cosmic Infrared Background
CMB	Cosmic Microwave Background
CME	Coronal Mass Injection
COBE	Cosmic Background Explorer
Co-I	Co-Investigator
Co-PI	Co-Principal Investigator
CRAF	Comet Rendezvous Asteroid Flyby (Mission)
CRIE	Cosmic Ray Isotope Experiment
CRRES	Combined Release and Radiation Effects Satellite
CSSA	Coarse Sun Sensor Assembly
CSSP	Committee on Solar and Space Physics
DBM	Data Base Manager
DE	Dynamics Explorer
DMR	Differential Microwave Radiometer
DSN	Deep Space Network
EHF	Extremely High Frequency
EHIC	Energetic Heavy Iron Experiment
EIRP	Equivalent Isotropically Radiated Power
EMC	Electromagnetic Compatibility
EMI	Electromagnetic Interference
EMU	Extra-vehicular Mobility Unit
EOL	End of Life
Eos	Earth Observing System
ESA	European Space Agency
ESTRACK	European Space Tracking (Network)
ETR	Eastern Test Range
EUV	Extreme Ultra-Violet (Wavelengths ~100–200 nm)
FAST	Fast Auroral Snapshot Explorer
FOV	Field Of View
GAS	Getaway Special
GGS	Global Geospace Study
GHz	Giga-Hertz
GMT	Greenwich Mean Time

Appendix C Acronyms

GRO	Gamma Ray Observatory
GSE	Ground Support Equipment
GSFC	Goddard Space Flight Center
GSTDN	Ground Spaceflight Tracking and Data Network
GTO	Geosynchronous Transfer Orbit
HST	Hubble Space Telescope
ICD	Interface Control Document
ICE	International Cometary Explorer
IMP-8	Interplanetary Monitoring Platform
IR	Infrared
ISAS	Institute of Space and Astronautical Science – Japanese
ISTP	International Solar-Terrestrial Physics (Program)
JPL	Jet Propulsion Laboratory
JSC	Johnson Space Center
KSC	Kennedy Space Center
LAP	Launch Assembly Plan
LED	Light Emitting Diode
LEO	Low Earth Orbit
LRR	Launch Readiness Review
Mbs	Megabits per Second
MCC	Mission Control Center
MELTER	Mesosphere-Lower Thermosphere Explorer
MHD	Magnetohydrodynamic
MMU	Manned Maneuvering Unit
MO&DA	Mission Operations and Data Analysis
MOU	Memorandum of Understanding
MSFC	Marshall Space Flight Center
NASCOM	NASA Communications Network
NGT	NASA Ground Terminal
nm	Nanometer
NOAA	National Oceanic and Atmospheric Administration
NOSS	National Oceanic Satellite System
NSSDC	National Space Science Data Center
OMS	Orbital Maneuvering Subsystem
OSL	Orbiting Solar Laboratory
OSSA	Office of Space Science and Applications
PAM	Payload Assist Module
PBL	Planetary Boundary Layer
PCU	Power Control Unit
PDMP	Project Data Management Plan
PDR	Preliminary Design Review
PI	Principal Investigator
POCC	Project Operations Control Center
POF	Pinhole Occulter Facility
pps	Pulses Per Second
RF	Radio Frequency
rms	Root Mean Square

Appendix C Acronyms

RTG	Radioisotope Thermoelectric Generator
SAMPEX	Solar, Anomalous, and Magnetospheric Particle Explorer
SAR	Synthetic Aperture Radar
SEI	Space Exploration Initiative
SHF	Super High Frequency
SIRD	Support Instrumentation Requirements Document
SMEX	Small-Class Explorer
S/N	Signal to Noise Ratio
SOHO	Solar and Heliospheric Observatory
SPAN	Space Physics Analysis Network
SPO	Solar Polar Orbiter
SPOT	Système Probatoire d'Observation de la Terre
SSCE	Space Station Cargo Element(s)
SSF	Space Station Freedom
STS	Space Transportation System
STTP	Solar-Terrestrial Theory Program
TBD	To Be Determined
TDRS	Tracking and Data Relay Satellite
TDRSS	Tracking and Data Relay Satellite System
TIR	Thermal Infrared
TIROS	TV Infrared Operational Satellite
TOF	Time of Flight
TSS	Tethered Satellite System
UTC	Universal Time Co-ordinated
VHF	Very High Frequency
VLF	Very Low Frequency
WFF	Wallops (Island) Flight Facility
Wind	Interplanetary Space Physics Mission

www.ingramcontent.com/pod-product-compliance
Lightning Source LLC
Chambersburg PA
CBHW081718170526
45167CB00009B/3617